ECOLOGICAL AND FISHERIES DEVELOPMENT IN WETLANDS

A Study of Chilka Lagoon

D. R. Khanna
S. Singh

2024

Daya Publications House®

A Division of

Astral International Pvt. Ltd.

New Delhi – 110 002

First Published, 1995

Reprinted, 2024

ISBN: 978-81-7035-135-1 (HB)

Published by : **Daya Publishing House®**
A Division of
Astral International Pvt. Ltd.
– ISO 9001:2015 Certified Company –
4736/23, Ansari Road, Darya Ganj
New Delhi-110 002
Ph. 011-43549197, 23278134
E-mail: info@astralint.com
Website: www.astralint.com

Dedicated
to my wife Manju
for her constant encouragement
given to me during the
writing of this book.

Preface

Wetlands represent a transition zone between terrestial and aquatic systems with diverse ecosystems such as lagoons, marshes, swamps and flood plains. The different wetlands vary widely in ecosystem attributes including characteristic biota, species richness, productivity and the amount of organic matter accumulating in the sediment. This, in turn, attracts a varied population of aquatic fauna and birds in great numbers. Several of them are resident, while a good may of them are migratory. Wetlands also act as very good habitat for development of nursery grounds of shell fishes, crabs and other fauna. Stabilisation of sediments is also an inherent values of the coastal wetlands.

The wetlands are vulnerable to the biodegradable and stable compounds from the land. Accumulation of suspended and particulate matter in the estuarine region is alarming, which hinder the free passage of saline water into the tidal creeks and thus reduce the salinity levels of the environment.

Rational use of the wetlands is above criticism, when the landuse primarily aims at beneficiary purposes and of economic value without hampering any ecosystem. However, the indiscrimate conversion of the wetlands need be checked, when the land can be best used for regeneration of renewable energy resources.

All these changes are presently operating in Chilka lagoon, the largest coastal wetland of the country, identified in Ramsar Convention.

Judging the importance of the coastal wetland, such as Chilka lagoon, several problems have been associated with them, which needs immediate attention for the conservation and management of the rapidly deteriorating wetland ecosystem.

Chilka, the largest lagoon of Bay of Bengal on the east coast of India, Orissa State, is situated between 19°.30'—19°57' N. latitude and 85°5'20"—85°29'20" E longitude has been identified as a wetland of International importance at the Ramsar Convention.

Noticeable changes have taken place in physical features of the lagoon comprising bathy-metry, salinity, siltation and rapid weed infestation in recent decades.

Added to all this, is the human intervention by way of over-exploitation of fishery resources, putting up mud walls/bundhs or enclosures by net for pen culture on the fringe and shallow areas of the lake on a massive scale by individual prawn farmers, primary fishermen cooperative society, corporate bodies by converting even traditional fishery sources of the lagoon for prawn farming. All these natural and artificial factors lead to decline of fisheries and degradation of the eco-system of the lake.

The lake is the source of wealth and contributes to the State Domestic Product, foreign exchange earnings, provides livelihood and sustenance to more than 1.5 lakhs of population living in the neighbourhood villages, attracts tourists.

Changes in the lease policy in respect of Chilka fisheries from time to time had a profound impact on the socio-economic dynamics of the fishermen vis-a-vis-non-fishermen population, eco-system and environment of the lake.

The size of population being limited earlier, the man-land ratio as well as per capita fishery resources of the lagoon were relatively high. But over the years, through increase in population and depletion of fishery resources, stress and strain entered into the lake area resulting in social dissensions and communal conflicts and clashes among fishermen and non-fishermen and between fishermen themselves.

In the quest of increase in revenue earnings, the State Revenue Department also made a departure from usual fisheries lease policy of Chilka lake during 1991, without allowing the users and participants of the lake fisheries to have their say and bringing the correct factual aspects of the matter into the light.

Increase of revenue by itself may not be a good ground for departure from lease policy pursued upto the end of 1991, without ascertaining the effect of the changed policy on fishermen of the lake and without paying any attention on the ecology of Chilka lagoon.

Against this back-drop, an attempt first of its kind in the present book has been made to review and analyse the recent changes of physical features of the lagoon together with ecological characteristics, flora and fauna, particularly fish and fisheries of the lagoon.

The present commercial activities of the lagoon, the change in land and water use and the changing socio-economic conditions of population living in and around the lagoon with the shift from usual fishery lease policy have been dealt with cautions for degradation of environment affecting ecology and fisheries due to illegal and rampant fish farming activities in the lagoon with possible consequences. The agencies, which are actively taking part in these unlawful activities for earning more and quick money have been identified also.

Efforts have also been made in suggesting proposals to mitigate the current ecological, fisheries and socio-economic problems of Chilka lagoon. Restoration and optimizing the brackishwater conditions by improving the salinity gradient, recruitment of prawn and fish juveniles, controlling the spreading of weeds; decreasing the silt load; identification and equitable distribution of fishery sources together with the conservation measures suggested in the communication can only save Chilka, the largest wetland of the country from rapid eutrophication leading to its extinction in future.

The author is grateful to Sri B.S. Das for typing out the manuscript in time.

Dr. K.P. Biswas

Contents

Introduction

The Chilka lake, the largest lagoon of India spreading across the State of Orissa is subjected to stress and strain of natural hydrological, biological, geophysical and socio-economic activities of the humans occupies the focal point of any discusslon on the wetland conservation ecology of India. The lake do not contribute significantly in the total fish production of the State; but became the most controversial issue due to the system earlier prevailed in granting all entitlements to eligible members of Fishing Community for undertaking specified fishing activity on certain designated fishing grounds of the lake.

Although entitlements are granted to the eligible fishermen in the lake, the content of these entitlement differ from fishing ground to ground and among different fishing methods. The control of access of rights to lake for fishing and the regulation of demographic pressure have not in any way led to lower the intensity of fishing effort in the fishing grounds. On the contrary, the situation is further complicated by the falling productivity of the lake and un-authorised occupation of fringe areas and unused fishing grounds of the lake for prawn farming.

Chilka lake of Orissa State is in transitional stage of natural evolution from brackishwater to freshwater as significant physical changes occurred in the lake in recent decades which are (a) water spread area of the lake from 906 Sq. Km. in 1915 was reduced to 790 Sq. Km. in 1989, (b) 10 fold increase in weed covered areas between 1973 and 1985, which estimated to be expanding at the rate of 14.6 Sq. Km. per year, (c) remarkably shallow areas getting shallower with load of silt brought away by rivers and rivulets during flood, and (d) decreasing salinity, particularly, in the northern sector of the lagoon.

Whereas, the earlier system enjoyed the legitimacy to the lagoon fishermen and contribute to the prevention of conflicts within these groups, the Chilka lake has now come under increasing pressure from outsiders to expand prawn culture activity in the lake, which have a risk of very serious effect on ecological degradation, particularly, in a confined environment, such as, Chilka lake. In fact the recent survey estimated at 25 per cent of the lake margin areas now being used for prawn culture un-authorisedly, which if get uncontrolled, will eventually affect the ability of the lagoon to sustain its natural fish production besides altering the lagoon environment since the current method of dyke constructions results in alienation and removal of productivity in shallow water areas, that are lost as wild fish habitat.

The social, economical system behind the uncontrolled expansion of prawn culture in Chilka was more from 1985, when higher prices of prawn led to commercialisation of prawn culture and encouraging the non-fishermen those living around the lake to enter into prawn culture. While lagoon fishermen who are to work hard to catch fish and getting very low price are remaining poor, the non-fishermen, as a middle man, take the profit. This resulted non-fishermen becoming stronger socially and economically and forcibly occupying and own the land. The entire fish economics of the lake, which was once controlled by the lagoon fishermen, is now under the custody of non-fishermen and the lagoon fishermen as fishing labourer getting poorer day by day. Legally the non-fishermen have no ownership either over any land, which is created due to sedimentation of lake water in which they are doing prawn culture. The land belongs to the lagoon fishermen, who are the lease holders to fish and hence the lagoon fishermen have the sole right over the lake water which they are enjoying from time immemorial.

But at present the non-fishermen either have forcibly occupied the land and water or have taken water rights on sub-lease basis by using their socio-economic pressure and some time physical muscle power to retain their hold over land and water and also to acquire more land.

The owner of the land and water of Chilka lake and law enforcing authority, the District Collector summarised the situation and their helplessness as follows:

"Entry of the non-fishermen into fishing through removal of social and psychological inhibition has resulted in conflict of interest between the fishermen and non-fishermen. The fishermen, who are a small minority in terms of numbers, want to hold on to the rights conferred on them by Government in it's 1959 and 1988 policy. On the other hand lured by the gold mine of prawn culture, the non-fishermen have forcibly occupied the Dians and Uthapanis leased out to the primaries and have been operating them inspite of constant conflict with their neighbours and in the face of still resistance from the Government. The Government have been reduced to the level of becoming a helpless spectator. In the scheme of things, the elected Board of Central Society has played a dubious role. Instead of reporting the matter to the Collector, to protect the rights of the fishermen and enforcing the lease conditions, they have encouraged large-scale encroachment of Dians and Uthapanis by the non-fishermen. Right now it has become a fate-accompli and it is not possible to do anything about it. The non-fishermen community have spent lakhs of rupees in constructing bunds in the Dians for converting them into prawn culture tanks and there will be nothing but blood-shed in case the Government tries to intervene to remove these constructions. Cases decided by the Court of Collector and also in the Honourable High Court of Orissa have remained as pieces of paper."

Serious consequences of the un-controlled prawn culture on the lagoon's environment is apprehended with the constant desire of grabing more and more grounds for prawn culture. The barrier of existing technology (not harmful to the lagoon's environment) will collapse soon to the advent of big industry stepping in the scene. In the competition of acquiring more and more land and to produce more and more prawn by constructing physical barrier and allowing the effluent rich organic waste water from the cultured area into the lake system, high environmental risk of

(i) the environmental effects of antibiotic drugs usage and contamination from materials such as, chemicals, which will be used for water and pond treatment;

(ii) alteration of the wild population gene-pool caused by pond shrimp escapes;
(iii) the possibility of shrimp pond operations to be sources for disease, which will infect the wild fish and prawns of the lake; and
(iv) the reduction of fish and prawn population of the lake because of the capture of brood stock or seed stock is not very far.

Critics make debate on the low production of commercially important prawn and fin fish from the lagoon and may argue on the fuller utilisation of the shallow areas of the lake through prawn farming, particularly, in the present state of environmental degradation of the lake. But considering the confinement and shallowness of the lake, the brackishwater environment is required to be maintained to the greater part of the lake to assist establishment of congenial environment for growth, survivality and production of economically important fishes and prawn and attract the ingress of more prawn and fin fish juveniles into the lake, without any obstruction due to physical barrier or environmental degradation of the lake and the connecting channel. By allowing juveniles of prawn and mullet and other important fishes from the nearby in shore areas of the Bay of Bengal into the lake freely, more fish production and income from the wild areas of the lake benefitting lagoon fishermen, will be achieved.

To decrease the disputes of entitlements to the fishing grounds of the lake, re-survey and re-identification of fishing areas, which are now in-accurately situated, is required. Re-establish the lease values based on the capability of the areas to produce fish and prawn together with rationalisation for distribution of the fishing grounds with complete ban on the interception of migration of both adult and juvenile prawn, fin fishes in connecting channel and Palur canal are necessary.

More than three hundred documents have been prepared on the lagoon by different authors from time to time with academic, faunistic, ecologistic and socio-economic interest.

Dozen of reports have also been prepared by different agencies from time to time highlighting the degradation of environmental condition and ecology of the water bodies, such as, progressive shrinkage of the water area, rapid weed infestations, siltation, with the apprehensions of future depletion of the capture fishery on which 85,000 fishermen of the lagoon depend.

On the other hand voice have been raised for destroying the Chilka ecology by allowing the fringe and jano area of the lagoon for prawn farming as a measure of enforcing highest economic activities not only for higher production through scientific culture but also for foreign exchange earnings.

But it is unfortunate that, no attention has been paid about the present condition of the lagoon and its subsequent impact on the fishermen, who sustain on the Lake. Once reported surface area of 2200 Sq. Km. has been shrunk into 790 Sq. Km. due to siltation and rapid weed infestations converting 40 per cent of the Jano fishing ground of the Lagoon more or less un-productive. The majority of fishermen population of the Lagoon, has no other income apart from fishing. When they are deprived off good productive fisheries source in the Lagoon, they find difficulty for a substitute earnings. Different economic survey among the fishery folk around Chilka reveal as many as.80 per cent of them are in debt atleast for some part of the year.

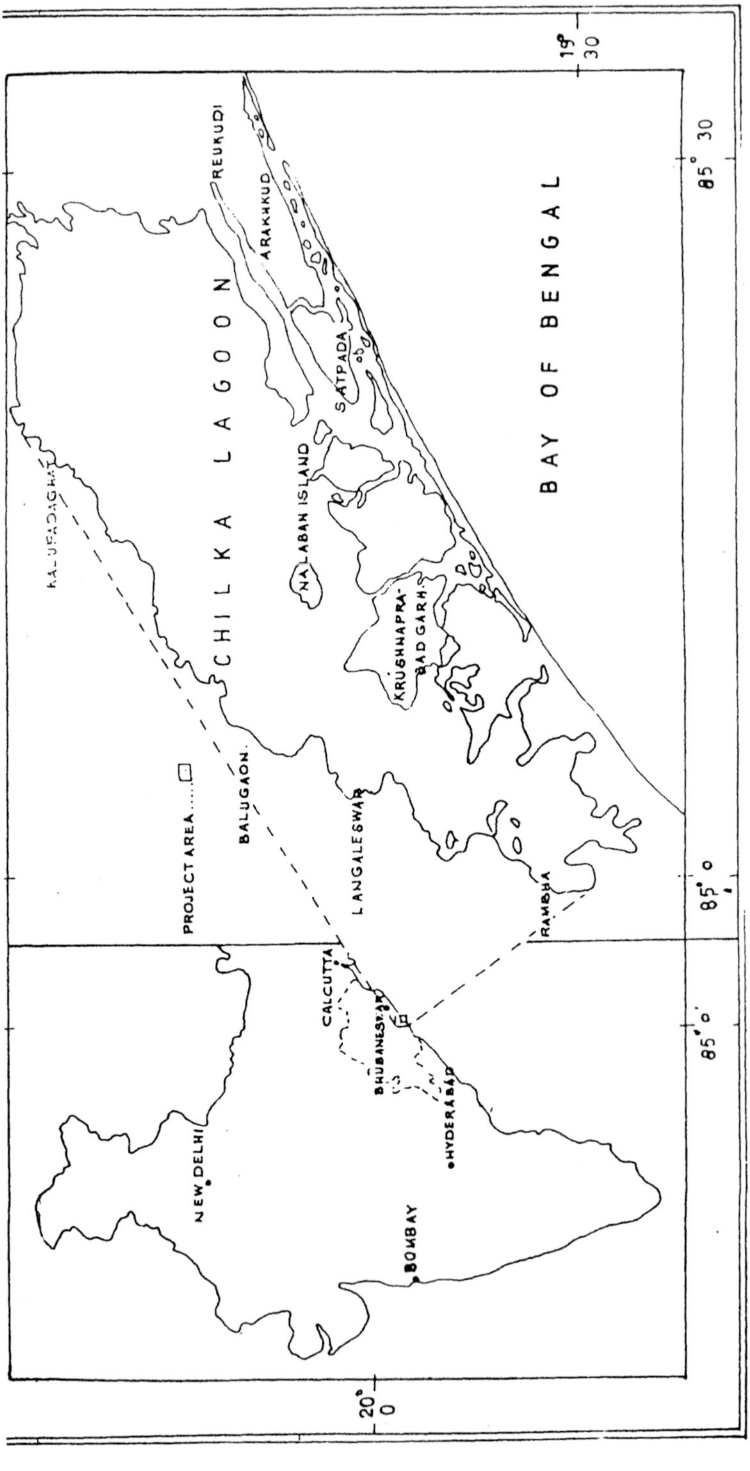

Location Map of Chilka Lagoon

Many of the P.F.C.S. are poor and even many of the members of the Society have little knowledge about economy and marketing. Most Primaries can not pay their members at the time when they come with their catch, because they are lack of capital. The conflict become more so, when some fishermen and non-fishermen take advantage of the poor Primary Society, convert Jano areas into prawn ponds illegally and live luxurious life, as can be seen from 2 storey-concrete house among the thatched houses and mud-walled houses. To improve the life of the fishermen, education is needed to make them aware of, what is happening in Chilka and the most urgent need to-day, is education in fisheries conservation, to make them understand the importance of migration of fish and prawns from the lake to the Bay of Bengal and vice versa, their rational exploitation and importance of marketing and maintaining the ecology of the lagoon.

In the present communication attempts have been made to understand the realistic condition of the fisheries of Chilka lagoon both in past and during present days so as to get in idea of the present situation and future fate of the lake fishery together with the present condition of the lagoon fishermen, who are enjoying the traditional fishing rights in Chilka lake from time immemorial.

1

Limnology

Chilka—A Lake or Lagoon?

The geomorphic, physiographic hydrographic and faunistic features of the coastal body of brackishwater spreading across the Ganjam and Puri districts of Orissa State, situated between latitudes 19°28' and 19°54' N and longitudes 85°05' and 85°38' conform to the characteristics attributed to a "typical" coastal lagoon (Phleger, 1981, Barnes, 1980). Rather the Chilka lagoon can be identified as Estuarine lagoon since it is fed with a number of river and rivulets; and also connected to the adjacent Bay of Bengal by a long narrow channel.

Genesis of Chilka Lagoon Basin

Chilka Lake, the largest brackishwater lagoon in Asia, like other lake or lagoon is not static. It reaches the late stages of natural evolution from marine to brackishwater to fresh water system. The rate of this transition has been accelerated due to land use practices, which increased erosion and sedimentation and reduced tidal exchange and flushing.

Significant physical changes are occurring in the lake; exhibiting:

(i) shrinkage of water spread area of the lake from 906 Sq. Km. in 1915 to 790 Sq. Km. in 1986.

(ii) ten-fold increase in weed-covered areas between 1973 and 1985, estimated to be expanding by 14.6 Sq. Km. per year.

(iii) being remarkable shallow and getting shallower (the minimum depth of 0.38-1.5 m. and maximum depth of 2.15-3.85 m. during high water in 1990-91 as against the maximum depth of 3.9 m).

(iv) highest salinity of 8-12 ppt. in the Southern end of the lake, which gradually decrease toward the north.

(v) hardly perceptible tidal effect in Rambha and at the mouth of Daya and Bhargabi rivers; and

(vi) dynamic nature of the shifting mouth of the lagoon over decades.

The potential loss of marine influences and lagoon characteristics due to evolution of freshwater system in Chilka lake, would likely cause significant decline in the diversity and value of the fisheries of the lake.

The physical changes which are occurring in the lake as a result of sedimentation—reduced depth and water area, reduced salinity, increased weeds, reduced diversity, low-value fish species, declining water quality and less productive aquatic environment, in one hand can be attributed to the natural cause, with equally important is the human interferences during last decade—the involvement of non-fishermen and growing conflict in the fisheries, the increased population dependent on the lakes resources, the disregard for conservation; especially the uncontrolled expansion of prawn culture into the lake and blockage of Palur canal unauthorisedly, which serves as entry path of prawn and fish juveniles into the lake, aggravates the physical decline of Chilka lake.

High world market prices for prawns has created such immense competition and combat over the resources.

Origin of the Chilka Lagoon

All evidences led to believe that the Chilka lagoon, which was once, a open bay and a part of the sea; become shallow by deposits from the mouth of the Mahanadi, and from silt carried up the Bay round the hills near Ganjam by the violent southernly winds of the monsoon and then entirely cut off by a spit, formed by the same agency, of sand drifted along the coast. The spit of sand formed across the mouth eventually separated the lagoon from the sea.

Formation of Chilka Lagoon

Lagoons are highly dynamic, ephemeral aquatic systems, which once formed, persists for hundreds to thousand years and are positively correlated with size.

When a portion of the sea is enclosed by offshore or longshore sand barriers, lagoons are usually formed. They are shallow systems with maximum depth rarely exceeding 10 m.; and because of this shallowness, lagoons are relatively easily converted into swamps, marshes and ultimately land by plant colonisation and encroachment.

Chilka is a classical tidal lagoon, created by a beach barrier berm that developed by the accretion of coastal sediments following the stabilization of sea levels some 3000-4000 years back and may persist for considerably longer than 1000 years. It differs from some lagoons because of the large influx of freshwater, particularly during the monsoon season, which gives the lagoon a brackishwater characteristic.

Three principal geomorphic features, which are the cause of creation and sustenance of Chilka lagoon are :

(i) the weathered hills that anchor the south-western limit of the lagoon;
(ii) the depth of the Mahanadi river to the north east; and
(iii) the barrier berm.

Two distributory channel of the Mahanadi River, the Daya and Bhargabi delivered sediments into the lake, infilled about 400 sq. km. of the lake leaving between 900 to 1200 Sq.Km. of water surface area at present. 735,000 cubic metre of silt are estimated to be loaded in the Chilka lake annually, through 42 point sources, 5 of which are larger rivers or streams.

An inlet of the lagoon is formed by the interaction between waves, current and sediments and the dynamic equilibrium develops at inlet depend on its cross-sectional area in relation to the tidal prism, or volume of water exchanged on a tidal cycle.

Besides receiving the benefit of the Mahanadi river flows in keeping the tidal inlet open, the tidal fluctuation in Chilka lake is about 0.2 m. during non-monsoon periods and upto 2 m during monsoon season, driven by a tide in the adjoining Bay of Bengal ranging from 0.9 to 2.4 m.

Occasionally the location of inlet migrates to a point where flow resistance in the channel is too great and hydraulic forces act to open a new inlet with a shorter connecting channel. The inlet of Chilka has been located in the same general area, within about 10 km. (Arkhakuda village) since at least 1914 and is at present connected to the lake through a 35 km. long channel.

Geographical Condition of the Lagoon

Chilka, a truly coastal lagoon as per definition accepted in UNESCO symposium on coastal lagoons held in Mexico in 1967, occupies the position between Mahanadi delta and Rushikulya estuary. The southern end of the lagoon is 18 Kms. away from Rushikulya mouth intervened by low lands, now partly used as salt pan. The northern side is characterised by number of swamps and water bodies, tidal creeks upto Brahmagiri. On the west the hill tract constitutes the watershed.

The lagoon has been formed due to the sea level changes and formation of barrier spit. The spit has grown for more than 60 Kms. near Palur hills, where the rocky promotory has resulted in wave refraction with variable width from 0.6-2 kms. The wind blown sands have formed frontal and back dunes on the spit with low interdunal stretches.

The inlet to the lagoon has varied in number and shifted its position several time during the recorded past. The opening has been generally shifting to the north and as many as three openings at a time has been recorded. The width of the opening has been recorded to be more than 1 Km. at times. In 1972 during a severe cyclone, two new mouths were created apart from the old one near Moto village. The old mouth became defunct in 1973 and for some years two mouths existed. The lone present opening, however, is about 300 metres wide opposite to Arakhakuda.

The shelf zone off Chilka coast is gentle for most part. There is a shallow sedimented zone extending far into the sea opposite Satpada, where an inlet existed and stabilised for long. The 15 metre depth line lies at 4.5 kms. from the coast at Palur hill and 12 kms. at Satpada. There is a submarine channel with corresponding ridge off present Chilka inlet. The shelf is steeper near Palur hills.

The long shore current is predominant from south to the north and northeast monsoon induced current from north to south is weaker. The net sediment transport is 0.7 m. cubic metre. The sediments transported by long shore current is thrown up on the beach by wave action. The coarse sand travel inside the channel, has been traced up to 4.5 kms. from the inlet after which the silt predominates. The wind speed is 5 to 10 kms. per hour for most part of the year. The maximum speed of 45 km./hour is experienced during monsoon. The cyclonic wind speed some time reach 100/km./hour.

The lagoon in the northern sector receives freshwater from 10 rivers, besides minor streams and nalas numbering 25 of which, the Daya, Bhargabi, Nuna Kansari and Mandakini, which join at the village of Balabhadrapur are major ones.

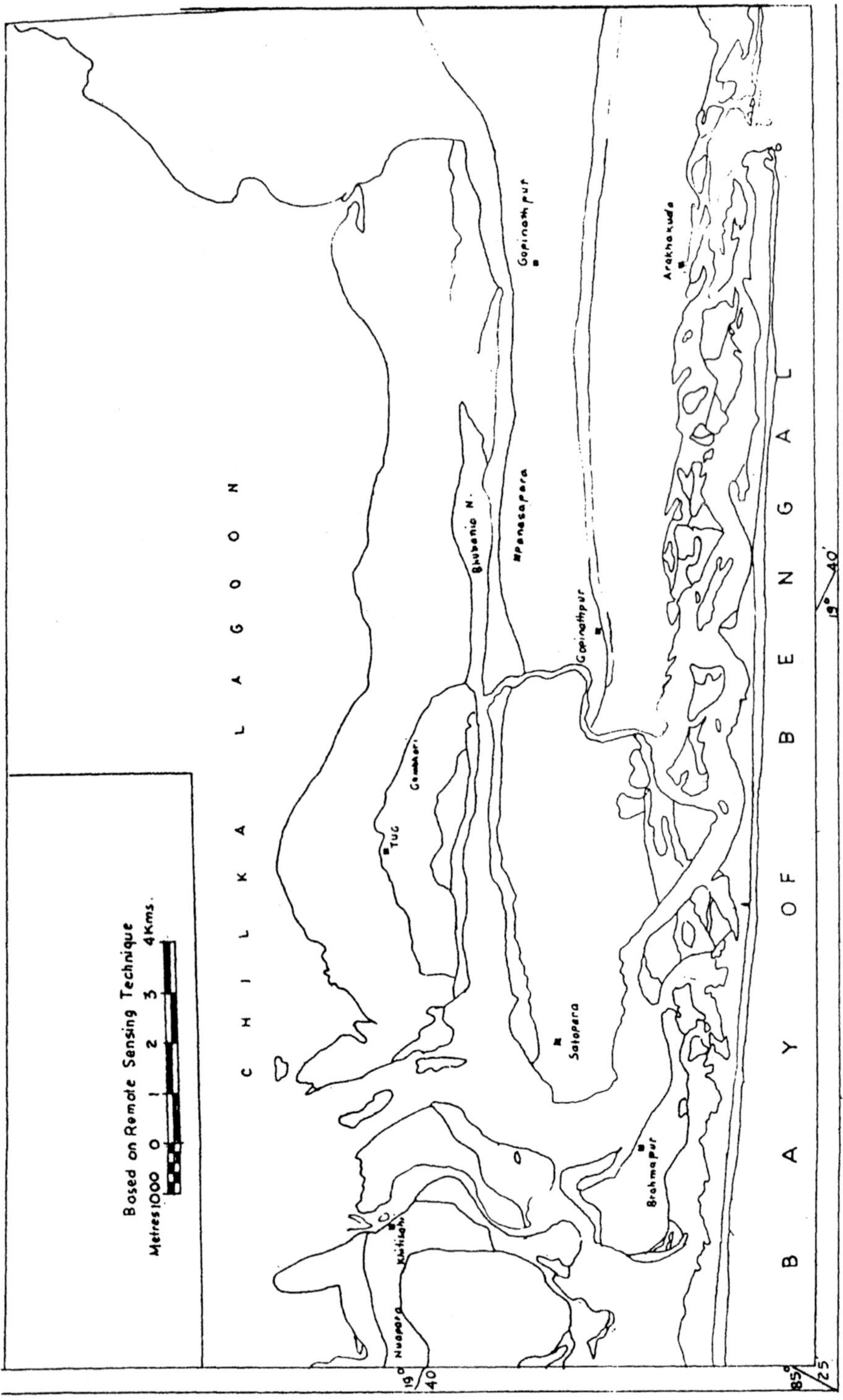

Magarmukh Area of Chilka Lagoon

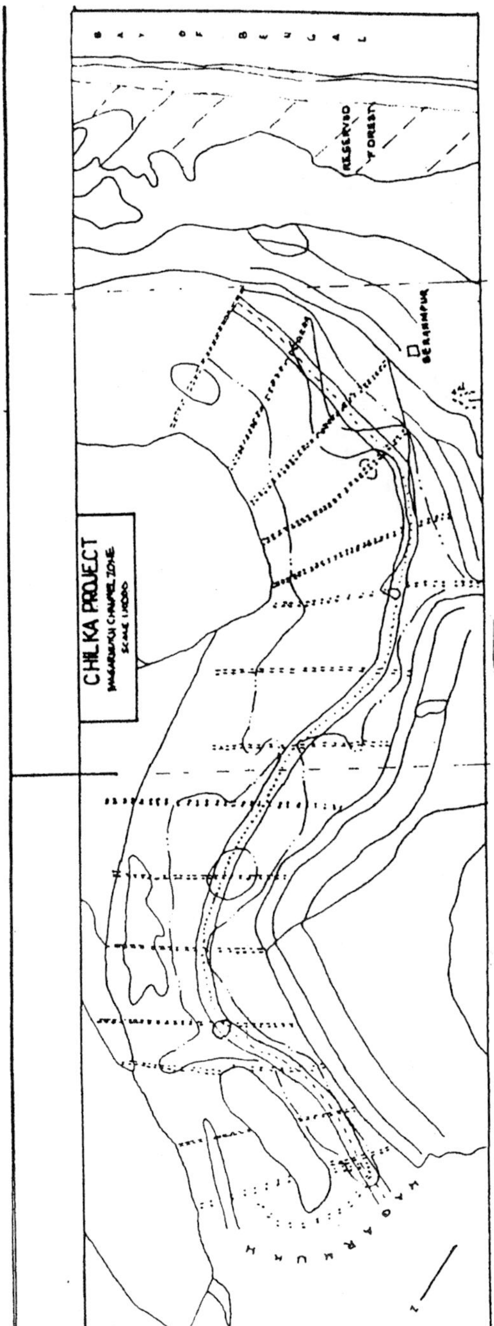

Outer Channel Connecting Lagoon to Bay of Bengal

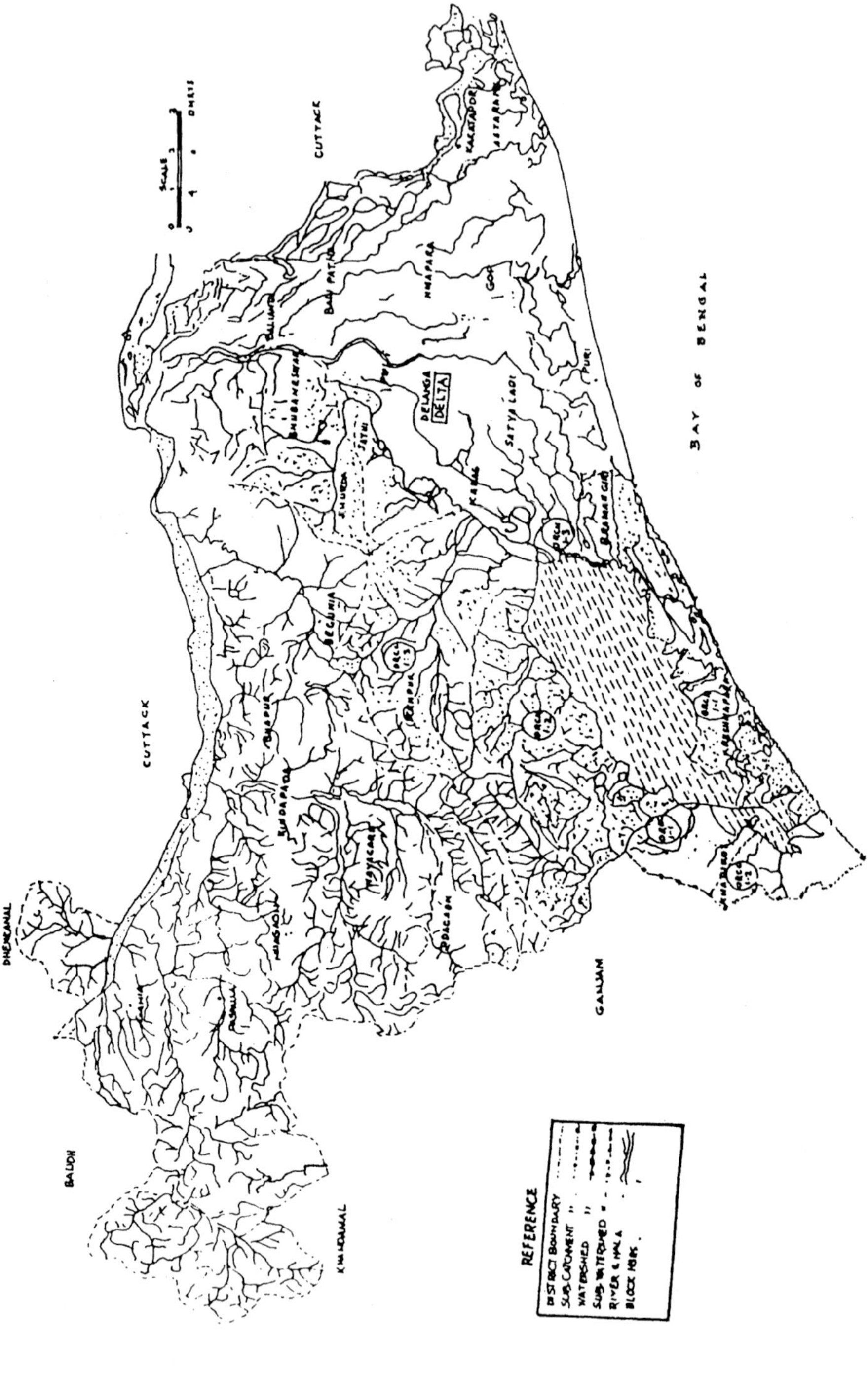

The Catchment Area of Chilka Lagoon

Discharge of flood water in the northern sector of the lagoon from the catchment area of 3560 sq.km. with average precipitation of 1.1 metre per annum accounts for 1760 million cum. The total annual freshwater flow into the lake has been estimated at 3,75,000 cusecs giving the average rise of the water in the lagoon by 2 metres, with maximum water spread area of 863 sq.km. carrying 1760 million cum. of water.

The capacity of the lagoon at the lowest water level in the summer become 966.97 million cum.

The Chilka water body accounts for 790 sq.kms., with a total of 4289 sq. kms. area delineated as the influence area of the lagoon has been classified as under:

(i)	Agriculture area mostly dry lands	2325 sq. kms.
(ii)	Reserve and protected forest	526 sq. kms.
(iii)	Area under plantation and permanent vegetation	717.7 sq. kms.
(iv)	Dense vegetation	73.6 sq. kms.
(v)	Sparee vegetation	226.2 sq. kms.
(vi)	Degraded forest (waste land)	246.39 sq. kms.
(vii)	Degraded vegetation	417.9 sq. kms.
(viii)	Culturable waste lands including undulating uplands	125.22 sq. kms.
(ix)	Unculturable waste land including rocky and steep sloping area	29.46 sq. kms.
(x)	Streams and rivers	16.65 sq. kms.
(xi)	Tanks and reservoirs	16.10 sq. kms.
(xii)	Wetland with dry mud and grassy mud flats	91.12 sq. kms.
(xiii)	Swamps with reeds	70.80 sq. kms.
(xiv)	Prawn culture ponds in outer periphery of the lagoon	3.875 sq.kms.
(xv)	Settlements	46.25 sq.kms.
(xvi)	National Highway	1.6 sq.kms.
(xvii)	Other roads	2.57 sq.kms.
(xviii)	Railway	1.17 sq.kms.

The lagoon is about 64 kms. on its longest axis and 20 kms. at its broadest on northern side. Palur canal has been constructed for allowing tidal influx from Rushikulya estuary into the lagoon.

2

Hydrography of the Lagoon

Zonation

The lagoon can be divided into 6 major zones considering the hydrographic parameters like, bathymetry, transparency, temperature, salinity, total alkalinity, free carbon-dioxide, pH, dissolved oxygen, inorganic nutrients like nitrates, nitrites, silicate, phosphate, total solids and total dissolved solids.

The 6 zones are :

(i) Barakudi-Bhusundapur zone;
(ii) Baradihi-Magarmukh zone;
(iii) Balugaon-Nalabana-Kathapenthaghat zone;
(iv) Balugaon-Kalijai zone;
(v) Rambha zone (Behind Kalijai hill to Rambha); and
(vi) Outer channel zone (Magarmukha to lake mouth).

Bathymetry

The bathymetry contours indicate the bottom configuration. More than half the area of the lagoon is under 2 metre depth. Area within one metre depth includes northern fringe between Kaluparaghat and Barakudi and around Nalabana. A trench area exists in the southern part of the lake between Kalijai and Rambha with 3 metre depth.

The present bathymetric data of September, 1987 indicates the maximum depth of 1.65 m. to 3.40 m. exist in Rambha zone and minimum depth of 0.51 to 1.6m. at Barakudi Bhusandapur zone. In 1993-94, the lagoon has become shallower, when the average depth of the lagoon varied from 0.6 to 1.55 m. in Bhusandapur Kaluparaghat, 1.8 to 2.1 m. in Kalijai; 0.9 to 1.62 m. in Balugaon and 0.65 to 1.29 m. in Rambha area.

Temperature

Varies considerably from sector to sector of the lagoon. The temperature range recorded during 1987-88 to 1990-91 is 17.0 to 33.5°C.

Diurnal variation of water and air temperatures earlier (1971-73) were between 20.2° to 29.5°C and 19.8° to 29.6°C respectively. Water temperature in the outer channel ranged from 23.16°C in December to 30.8°C in June during the period.

Highest temperature range has been recorded during the peak of the summer (16.0°-35.2°C); while the lowest during flood season.

Rambha zone has higher temperature value than that of other zones of the lagoon, because of being enclosed by surrounding hills and land with insignificant

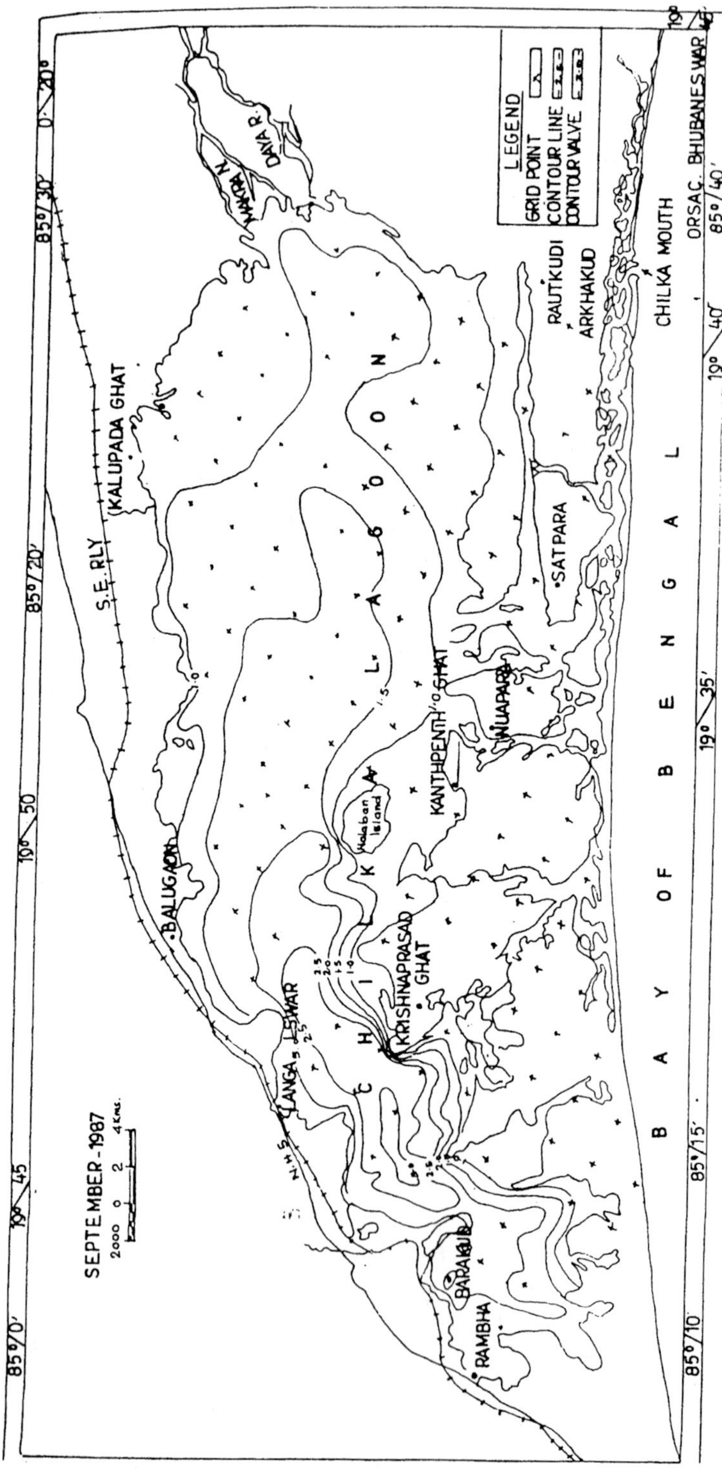

Bathymetry of Chilka Lagoon

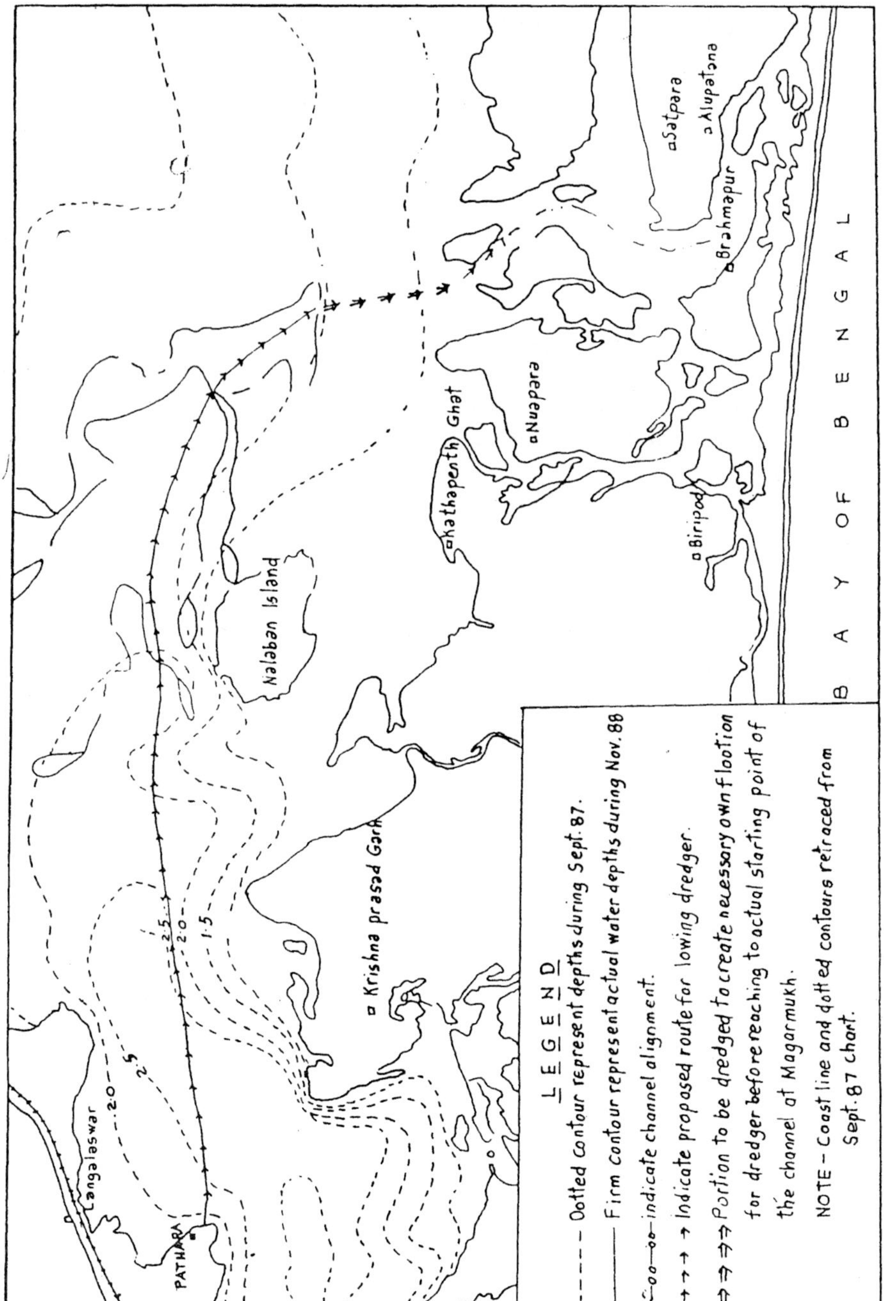

Chilka Lagoon Showing Nalabana Area

exchange of water with its adjacent central zone. The lowest temperature recorded in Barakudi-Bhusandapur zone are due to freshwater influx from Daya river. Surface bottom temperature differences are not significant in the lake. Surface temperature of the lagoon now range between 24.5° to 35.5°C with diurnal fluctuations of 1-2°C.

Salinity

Salinity of the lagoon is influenced by the river discharge in flood, season as well as, sea water influx in Summer and showed cyclic changes. The lagoon is brackish, over most of list area and varies from freshwater (0 ppt.) adjacent Daya river discharge to a higher saline level of 43.7 parts per thousand (ppt.) in the connecting channel during dry period. The variation is along north-south gradient.

Four different types of environment depending upon their salinity conditions prevail within the lagoon which are :

(i) Freshwater dominated zone (Bhusandapur-Daya, Bhargavi-Confluence-Bhudhakera-Kumandol-Tua area)

(ii) Brackishwater zone (Kumandol-Tua-Jalia hill-Parikud)

(iii) Sea water dominated zone (Behind Kalijai hill to Rambha)

(iv) Hyper-saline zone (Magarmukha to lake mouth)

The Brackishwater zone of a lagoon is defined, an area containing salinity between 5-20 parts per thousand (ppt).

High saline water prevails in the Rambha zone and it gradually decreases towards northern end of the lagoon. The high salinities in the Rambha zone is due to the semi-enclosed position of this part of the lagoon.and high evaporation rate. The lower salinities in Barakudi-Bhusandapur zone are due to freshwater discharge from Mahanadi river system. In summer the Barakudi-Bhusandapur area maintains its freshwater nature.

Salinity distribution contour of the lagoon show a tongue like intrusion of the Balugaon-Nalabana-Kathapenthaghat zone of the lagoon into the northern part and this area remains as a boundary between the more saline Balugaon-Nalabana-Kathapenthaghat zone and low saline Barakudi-Bhusandapur zone. This boundary develops after the cessation of monsoon and normally remains till the next flood.

The salinity of the lagoon starts rising from November-December and reaches high values in April-June. In 1990-91, the year of high precipitation, the salinity of the lagoon water dropped and remained less than 3 ppt. for

11 months in Barakudi-Bhusandapur zone.

6 months in Balugaon zone.

4 months in Kalijai zone.

1 month in Rambha zone.

Surface-Bottom salinity variation exists throughout the lagoon, but it is pronounced in Barakudi-Bhusandapur zone and maximum in outer channel as much as 5 ppt.

The salinity values recorded in 1990-91 shows 2.8-10.6 ppt. at Rambha zone, freshwater to 1.6 ppt. at Barakudi-Bhusandapur zone, and 4.43-31.7 ppt. at the connecting channel. In 1993-94, however, the salinity range in Rambha zone was observed to be 5.6-10.7 ppt; while that of Barakudi-Bhusandapur zone and connecting channel as 0.6-3.8 ppt. and 0.2-35.6 ppt. respectively. The salinity of boundary zone (Balugaon-Nalabana-Kathapenthaghat zone) was between 4.5-13.7 ppt. during the year. The water throughout the length of the channel is almost fresh upto lake mouth during monsoon and post monsoon season, when a strong unidirectional current created by large amount of freshwater discharge from Mahanadi river systems in the northern sector of the lagoon flows out of the lake through this channel. The channel is filled with sea water during dry season.

The progressive reduction of average salinity has been observed by the recent spread of freshwater macrophytes and detection of freshwater leaches in the northern sector of the lagoon. The annual mean salinity of whole of the lagoon dropped from 17.5 ppt. in 1957-58 to the lowest 4 ppt. during 1990-91 and thereafter recovered to 10.9 ppt. in 1993-94, which can be observed from the Table 2.1.

Table 2.1: Average Salinity of Chilka Waters for Four Major Zones During Past 31 Years (1963-64 to 1993-94)

Year	*Annual mean salinity of whole lagoon in ppt.*	*Trend values by 5 yearly moving average method*	*Annual average of sectoral salinity of Chilka lagoon in ppt.*			
			Northern Sector	*Central Sector*	*Southern Sector*	*Connecting Channel to Bay of Bengal*
57-58	17.5	—	13.7	15.5	19.1	21.8
60-61	13.7	—	10.9	11.7	14.1	17.8
63-64	9.5	—	5.9	7.1	9.9	15.2
64-65	12.4	12.1	10.6	13.1	11.5	14.4
66-67	12.2	12.1	8.1	12.5	14.1	14.1
67-68	12.7	12.3	8.3	12.7	14.5	15.3
68-69	13.9	11.6	11.1	11.9	14.0	18.4
71-72	10.5	10.8	6.4	7.8	10.9	16.9
72-73	9.0	10.2	5.8	7.5	8.7	13.9
73-74	8.1	8.7	4.2	7.3	8.5	12.5
85-86	9.8	8.3	6.0	7.0	8.0	8.4
87-88	6.4	8.1	1.5	8.9	8.4	6.7
88-89	8.2	7.3	1.5	9.0	10.5	11.8
89-90	6.7	6.5	0.7	7.6	12.3	6.2
90-91	4.0	6.9	0.5	2.6	6.8	6.0
91-92	5.8	7.4	1.1	7.5	7.7	6.8
92-93	8.5	—	2.1	7.4	10.1	14.3
93-94	10.9	—	1.7	8.1	7.8	25.9

The trend is conspicuous in all the sectors of the lagoon i.e. freshwater dominated zone, brackishwater zone, sea water dominated zone, hyper-saline zone represented in northern sector, central sector, southern sector and connecting channel to Bay of Bengal. The trend value of 12.1 ppt (5 yearly moving average) of mean salinity of whole lagoon from 64-65 exhibited a down-ward trend till 1990-91

(6.9 ppt) with marginal variation between different years. A slight increase in the trend value (7.4 ppt.) is however, noticed from 1991-92 to be confirmed in subsequent years.

The monthly salinity fluctuation of different zones of Chilka lagoon during last 7 years i.e. from 1987-88 to 1993-94 exhibit, the drop of salinity in all sectors of the lagoon during monsoon and post monsoon months with almost fresh water (0.2 to 0.6 ppt.) condition in the freshwater dominated zone and the connecting channel to the Bay of Bengal (hyper-saline zone) when large amount of freshwater discharge of river Daya and Bhargabi fall in the northern sector of the lagoon and flow out of the lake through the connecting channel. The rise of salinity in all sectors of the lagoon was noticed from December onward and remain maximum (3.5 to 31.6 ppt.) between the months of April to June i.e. prior to the monsoon as observed from the Table 2.2.

Table 2.2: Monthly Average (1987-88 to 1993-94) Salinity in ppt. of Different Zones of Chilka Lagoon

Month	*Average salinity in ppt. of different zones*			
	Northern	*Central*	*Southern*	*Connecting channel to Bay of Bengal*
April	2.4	6.5	9.0	31.3
May	3.5	9.6	8.9	31.6
June	2.7	12.2	9.9	25.0
July	0.7	11.2	10.6	22.8
August	0.3	7.7	10.1	6.8
Sept.	0.2	5.2	8.4	0.6
Oct.	0.4	5.4	8.7	10.2
Nov.	0.4	5.4	8.2	14.4
Dec.	0.5	5.7	8.3	15.3
January	0.9	6.3	8.5	14.6
February	0.7	5.8	8.6	7.9
March	1.2	5.9	8.9	12.5

The physico-chemical features of the connecting channel from Chilka lagoon to Bay of Bengal has paramount importance, as it serves the migratory path for the brood fish and prawn to the adjacent Bay of Bengal for breeding. The channel also act as the passage for recruitment of post-larvae and juveniles of prawn and fish for autostocking the lagoon. In 1963-64, the salinity of the connecting channel, was between 24.5 to 27.3 ppt. during summer (May); between 1.6 to 2.1 ppt. in monsoon (September) and 4.39 to 19.8 ppt. in winter (January). After 9 years (1972-73), the salinity regime of the connecting channel was between 34.73 to 35.13 in Summer (May); 0.2 to 4.6 ppt. in monsoon (September) and 6.7 to 33.3 in winter (January). During 1993-94 the salinity pattern of the connecting channel was observed to be 27 to 27.5 ppt. in summer (May); 0.2 to 2.2 ppt. in monsoon (September) and 3.4 to 28.4 ppt. in the month of January. The physico-chemical features and the salinity of channel water is most important as it plays an important role in recruitment of prawn and fish. Eighty eight juveniles of 12 to 18 mm. size were found to have been recruited per metre cross sectional area into the lagoon in an hour at a water salinity of 19.8 ppt.

Transparency

The colour of the lake water are transparent, turbid, green and yellowish green. The green colour is possibly due to moderate amount of dissolved matter and suspended particles, such as, fine clay. Yellow colour is due to the presence of large amount of organic substances.

The annual average secchi disc transparency of the lagoon water ranged from 0.003 to 8.5. The Rambha zone is found to be more transparent (1.51-5.31) than the other zones due to its semi-enclosed situation and less drainage. The Barakudi-Bhusandapur area is usually turbid due to sediment load carried through river discharges. During monsoon period, the transparency becomes very low throughout the lagoon due to entry of silt laden rain water from land drainage.

Dissolved Oxygen

Dissolved oxygen is a vital factor of the lagoon ecosystem. The lagoon water is usually well oxygenated almost throughout the year; and it ranged from 5.3mg./l to 11.2mg/l. Relatively higher oxygen value (7.6-11.2mg/l) has been recorded from Baradihi-Magarmukh area. Extremely high and lower values in southern area and in channel are due to thick vegetation and poor mixing process respectively.

Higher concentrations of dissolved oxygen (7.6-11.2 mg/l) at Barakudi-Bhusandapur zone and Baradihi-Magarmukh zone are due to the production of oxygen by photosynthesis of the dense plants and weeds found in the same region of the lake alongwith continuous flow of freshwater from Daya and Bhargabi rendering minimum water temperature and holding more dissolved oxygen. Low concentrations of dissolved oxygen are marked in Rambha, Balugaon channel and inlet area with high salinity.

Surface-bottom difference is appreciable in the southern sector due to improper mixing. Dissolved oxygen values of Chilka water have recently (1991) been estimated between 3.3 to 18.9 mg./L. During 1993-94 the dissolved oxygen of different sectors were estimated as:

Northern Sector	—	3.0 to 4.8 ppm.
Central Sector	—	4.8 to 5.8 ppm.
Southern Sector	—	3.8 to 5.8 ppm.
Connecting channel	—	4.8 to 5.8 ppm.

pH

Most of the lagoon have a pH range of 6 to 9.6 pH value increase from outer channel and central sector to north and south of the lagoon. The maximum pH, 9.6 was recorded at Rambha zone and the minimum pH 6.8 at channel area.

The variation in pH can be attributed to the influence of flood water influx from rivers and intrusion of sea water. Moderately alkaline condition of pH 6 to 9 (Optimum for aquatic life) prevails in the lagoon with highest alkaline value in the southern part near Rambha.

For propagation of fish the pH range of 6.8 to 9.6 in Chilka lagoon is found to be optimum.

Total Alkalinity

The total alkalinity of the lagoon was in the range of 20.0 mg./l to 260 mg/l, which now estimated to be between 55 mg./l to 125 mg/l, with a relatively higher value of 85-115 mg./l in Rambha zone. The present variation in total alkalinity follows the trend of salinity change.

Free Carbon-dioxide

Except Baradihi-Magarmukh zone and outer channel, where mean value of free carbon-dioxide are higher (2mg/l), the values in other areas are estimated to be in traces.

Nutrients

The nutrients in the lagoon are mainly supplied by the sediments and also supplemented by sea water.

(a)Silica ranges from 1-8 ppm.
(b)Phosphates from traces to 0.4 ppm.
(c)Nitrates account for 10-60 ppm.
(d)Ammonia varies from traces to 4 ppm.

Since diatoms require silica for shell formation and they form a very important constituent of plankton of the lagoon; their availability at higher concentration is conducive to phytoplankton (diatom) growth and satisfactory primary production.

Higher phosphate values and ammonia-nitrogen have been recorded from the channel and nearby Magarmukh area. Higher nitrate rates (30-60 ppm) was noticed in Balugãon Nalabana-Kathapenthaghat zone, while lower values (10-30 ppm) from Rambha zone.

Precipitation

With an average precipitation of 1.1 metres per annum in the catchment area, the annual rainfall recharge of the lagoon is of the order of 1760 million cum., allowing 15 per cent, infiltration, 5 per cent over land depression storage and 25 per cent evaporation. Besides the rain water interception in the lagoon is of the order of 870 m.cum. Calculating the evaporation of 1286 million cum of water from the lagoon, the net annual rainfall recharge of the lagoon is of the order of 1344 million cum.

Evaporation

Evaporation from the Chilka lagoon is more during summer than in winter, which indicate that the lagoon water is subjected to large seasonal variations unlike open ocean, the maximum evaporation is at the time of maximum temperature.

Lagoon mouth experience lower evaporation in summer than in winter as compared to the main lagoon due to the influence of the sea.

Maximum evaporation during afternoon hours of the day and minimum during night time have been observed.

Currents

Vertical and returning currents in Chilka lagoon are not encountered due to the shallowness of the lagoon. Horizontal current generated by wind occur in main lagoon. Due to the in-flow from the adjacent Bay of Bengal into the lagoon through the lake mouth during summer, tidal currents occur in the outer channel and at some places near Satpada. During monsoon and post monsoon, huge quantity of river water flow from Mahanadi river system into the lagoon. The current in the northern part are mainly due to this river water outflow from the lagoon into the sea and they overcome the tidal currents in the connecting channel. The recent study made in the lagoon mouth and the connecting channel during January to April 1992, indicates that the direction of current were along with direction of channel without any cross current. The main stream from connecting channel to Alupatna (27 km. from lagoon mouth) is along the land area on the sea side, when measured with a direct reading current meter in the main stream. The current strength is less in the village side of the connecting channel (near opposite bank). The pre-dredging current data obtained from lagoon mouth and connecting channel during January to February 1992 were:

	Rate in meters/min.	
	at flood	*at ebb*
Mouth	6	7
Arakhakuda	4.5	6
Sanpatna	5	10
Mirijapur	2.5	7.5
Alupatna	1	5
Satpada	0	1

The post dredging data after excavation of 82,500 cum of sand from the mouth of the lagoon of which 39,000 cum excavated from the central area and 43,000 cum. excavated from western area, the rate of current have improved along the channel from mouth to Satpada to the extent given below:

	Rate in meters/ min.	
	at flood	*at ebb*
Mouth	60	48
Arakhakuda	14	10
Sanpatna	7	10
Mirijapur	5	9
Alupatna	4	8
Satpada	3	5

The current at the mouth is considerably stronger in post dredging period causing erosion on northern coast.

Tide

At lagoon mouth the tidal graph is regular. The cycle of high and low waters and that of spring and neap tides is well defined. The cycle of spring and neap tides is well defined in the outer channel, namely at Arakhakuda (7.5 kms. from mouth), Sanpatna (13 kms. from mouth), Mirijapur (19 kms. from mouth) and Satpada (35 kms. from mouth). The time difference of high and low waters at lagoon mouth and at other 4 above mentioned places are:

Station	*Time Differences*
Chilka mouth	0
Arakhakuda	+1 hrs.
Sanpatna	+2 hrs.
Mirijapur	+3 hrs.
Satpada	+4 hrs.

The force of tidal stream diminishes with the distance from Chilka mouth.

Observation on 20 days cycle in April 1992, round the clock, at each of the following stations, both during neap and spring tides reveals the difference between high and low water as below:

Area	*Range (Mtrs.)*	
	Maximum	*Minimum*
Mouth	1.29	0.62
Arakhakuda	0.90	0.47
Sanpatna	0.70	0.45
Mirijapur	0.40	0.15
Satpada	0.22	0.05

When compared with the tidal ranges obtained during January/February '92 i.e. during pre-excavation period, a slight increase in high waters have been noticed after excavation which can be observed from the following pre-excavation data.

Area	*Range (Mtrs.)*
	Maximum
Mouth	1.22
Arakhakuda	0.72
Sanpatna	0.54
Mirijapur	0.39
Satpada	0.12

After excavation there appears to be no change in tidal height in lagoon mouth though the tidal ranges have shown some change with the distance from the mouth.

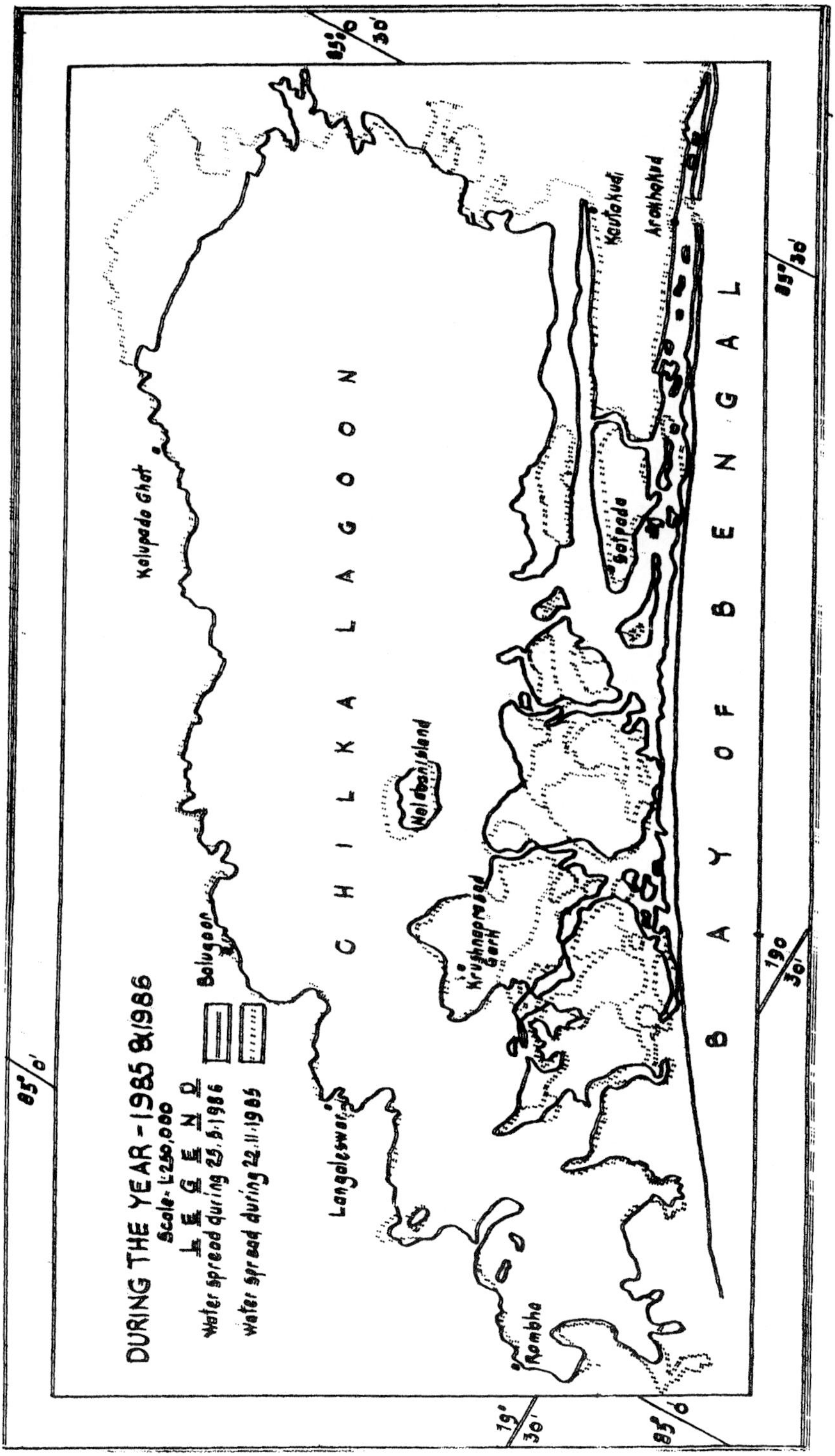

Waterspread Area of Chilka Lagoon

The tidal graph at Satpada is erratic and does not show any regularity in rise and fall. This indicates that the flow of water in the area is effected by forces other than tidal rise and fall. However, in general the tidal effect is indicated as the neap and spring tides are well defined on the graph though the range of tide is nearly the same. With the rise in the limit of high water the limit of low water also rises and vice versa. The period of ebb (7/8 hrs. approx.) is more than period of flood (5/6 hrs. approx.)

The tide graph of Mirijapur is more stable than that of Satpada though at times it does indicate forces other than tidal stream acting on heights and times of tides specially low water. The period of ebb is more than period of flood specially at spring tide.

At Sanpatna, the tidal wave is better defined. The period of ebb is 7 hrs. where as flood five hrs. on an average. The tidal cycle of spring and neaps is also more defined.

At Arkhakuda, the period of ebb is still more than the period of flood. The tidal cycle of spring and neaps is much better defined. There seems to be very little effect of forces other than tidal stream on heights and periods of tidal rise and fall.

Sedimentation

The sedimentation and shrinkage of the Chilka lagoon is prime concern. The Chilka is being silted up at an accelerated rate. The water has covered some 2200 sq. kms. of the lagoon earlier, compared with 1165 sq. km. after the monsoon in mid twenties computed from topo-sheets and from to day's 916 sq.km. obtained from recent remote sensing data. Thus in a time span of 60-65 years, there has been decrease in area of about 249 sq.km. the silting rate of the periphery being 3.5 to 4 sq.km. per year which is very high.

A study of the satellite imageries indicate that 46 sq. kms. area have been silted up and presently under swampy condition, where seasonal agriculture is being undertaken in part. A large part of it is covered with water hyacinth and reed grass located mostly on the northern fringe of the lagoon occurring between Bhusandapur and Barakudi.

The sedimentation is caused by over land flow from the western catchment as well as by the discharge of Daya-Bhargabi-Nuna-Mandakini rivers carrying a silt load to a tune of 13 million tonnes per annum. The sedimentation is also accelerated by two other processes. Firstly, by prolific aquatic vegetation in marginal areas in northern part arresting the silting up of the shore region, and secondly by constraining the low water margin to retain for ponding monsoon water and using as freshwater for fishery ponds by the local inhabitants. These ponds act as receptacle for the sedimentation and permanently make the area free from water spread. Year after year the process is repeated and the reclaimed land margin advances into the lagoon.

The sand carried by long shore current and wave action on shore enters the channel during high tide and gets deposited in the ebb period. Thus a group of islands have been formed by silt deposits brought by the sea water influx from Bay of Bengal. The channels have been silted up and large number of channel islands are formed reducing the passage as well as raising the floor. The formation and growth of the barrier spit is greatly helped by wind action and the dunes are formed. These dunes migrated towards the lagoon with passage of time.

Another cause of enhanced sedimentation seem to be purely a physico-chemical process. Due to silting of the mouth or the opening to the sea, the salinity of the lagoon has been in a decreasing trend. The silt load from the Daya-Bhargavi, the two main rivers discharging their load to the lake in the North-Western side, which was ultimately getting saline water from lake, to get their silt load dispersed or remain in a floculated condition due to absorption of sodium ion, is not getting the same saline environment now. As a result, of this, the sediments are now settling in the Chilka bed itself instead of being pushed through the lake to reach the main sea. Moreover, the sub-merged as well as the roots of the floating aquatic weeds are congenial for co-agulation of electro-statically charged clay colloids. One can pull up any floating weed and see how much of clay remain absorbed to the root system. It can be well postulated, that the roots of the floating weeds are 'charged' bodies which offer scope to the silt or clay in colloidal form to adhere to the roots and ultimately fall down insitu to silt up the lake. The same mechanism seems to be applicable to the leaf surface of the submerged weeds. They apart from behaving as per the above mechanism, offer a surface at a depth much less than the bottom of the lake, where the silt quickly gets deposited.

Thus it has been observed that the bed of the Chilka lagoon is being gradually raised by the silt deposit brought in by the rivers and it is probable that in time the low mud flats, which are pushing their way southward from the mouths of these rivers will extend over the whole lagoon within a course of three to four hundred years.

Nature of Bottom

The entire northern sector consists of mud due to continued deposition of silt during flood. The middle portion of central and southern sector of lagoon also have muddy bottom. The western shore or southern-central sector are composed of clay with an admixture of organic debris. The sand bed are predominant in eastern shore of southern and central sectors. The bottom of connecting channel is composed of sand only.

The region around Island like Kalijai, Badakud, Sankud, consists of gravel mixed with sand and clay.

Lagoon bottoms are covered by soft sediments, with a particle size gradient of small sized sediments (silt) prevalent in the freshwater-dominated and brackish zones and coarse-grained sands prevalent near the mouth. Towards the lagoon mouth, more vigorous water movement keeps silts in suspension and renders water conditions turbid during periods of high wind. Sediments usually accumulate in lagoons; the sedimentation rate (which is increased by submerged and emergent macrophytes) reflects the net difference between the rate of riverine sediment input minus the rate of resuspension and outflow. Panda (1988) classified the surficial sediments of Chilka Lake as sand, silt and clay based on grain-size distribution. The sediments of the outer channel and also the areas close to the western margin and rivermouths contained higher sand percentages than the other parts of the lagoon system. The percentage of clay material increased in a lakeward direction in all three sectors of the main body of the lagoon. There was a higher percentage of organic carbon in the sediment of the northern sector, which may be associated with macrophyte decay in this region of the lagoon (Panda 1988).

3

Ecological Characteristics of the Lagoon

Relevant features of the ecosystem directly supporting or contributing to fisheries and aquaculture production of Chilka lagoon from lower trophic levels to commercial aquatic resources have been indicated.

High nutrient concentrations are often present in the lagoons as a result of both riverine nutrient inputs and effective nutrient recycling between the sediments and the water column.

Greater concentration and supply of nitrates in Barakudi-Bhusandapur area are due to the land use practices surrounding the lake. Nitrate-nitrogen accelerates algal growth for which algal population is more in Barakudi-Bhusandapur area. The higher concentrations of nitrates and phosphates as well as low salinity restricted to this zone signifies the rich growth and abundance of freshwater plants and weeds. Nitrate is also increased due to the abundance of Oxygen formed in this zone.

Phosphorus in the lagoon occurs in both organic and inorganic forms. The main inputs of phosphorus into the lagoon are from inflowing rivers like Daya and Bhargabi. Phosphates are contained in plants and bones of animals. Because of higher concentration of phosphorus in Barakudi-Bhusandapur zone, algal blooms are noticed with higher fish productivity.

Concentrations of some of the heavy metals from Chilka water were determined as:

(i)	Zinc	0.025-0.19	ppm	with higher concentrations near Rambha area.
(ii)	Iron	0.12-0.32	ppm	with higher concentrations from Kaluparaghat, Balugaon, Nalabana, mouth of river Daya, Island complex of Sanakud, Badakuda and Samal.
(iii)	Copper	0.02-0.04	ppm	with higher concentrations from Balugaon-Nalabana.
(iv)	Mercury	0.8-5.6	ppm	with maximum concentration from Palur canal.

Lagoons are often highly productive aquatic environment compared to other aquatic systems. The following features provide lagoons with the distinctive characteristics:

- the high degree of shelter from tidal and current action;
- the relatively stable salinity gradients;
- the soft mud and/or sand substrates;

— the well-mixed nature of the water column through wind action;
— extreme shallowness;
— organic richness;
— rapidity with which they change (over geological time scales):
— in climates with seasonal rainfall, and where major inputs of freshwater exist, a pronounced seasonal variation in salinity and/or water level.

In comparison with estuaries, contributions of phytoplankton and submerged macrophytes in lagoons are more important in production processes. Most of the production is consumed within the system, and there is less export of nutrients and organic material due to the closed nature of lagoons and of the unimportance of tidal fluxes. Carbon sources include phytoplankton, benthic and epiphytic algae and detritus derived from macrophytes. The latter detrital source is especially important as a souce of carbon. The pond weed, *Potamogeton pectinatus* decays very rapidly within the lagoon environment. Studies by Howard-Williams and Davies (1979, cited by Barnes 1980) suggest that under environmental conditions of 15-26°C temperature and 5-11 ppt salinity, most nutrient release from *Potamogeton* occurred during its first week of decay, and decaying processes are largely completed within 128 days. Within lagoons, detrital enrichment via. bacterial heterotrophs is the dominant trophic pathway supplying energy to biological consumers.

Most consumers are thought to acquire detritus benthic algae and epiphytes in an indiscriminate fashion via. deposit feed and/or browsing. Among vertebrates within lagoons (both birds and fish) most species are opportunistic omnivores or carnivores.

In summary, lagoons are extremely productive environments due largely to high nutrient inputs from surrounding land drainages, as well as efficient nutrient re-cycling. This high productivity supports lagoons fisheries for both fish and shellfish. Lagoons are ephemeral environments (on geological time scales) evolving rapidly into other types of semi-aquatic, habitats (marshes, swamps). Simultaneous with this succession is a gradual shift from high salinity conditions to freshwater Human activities within lagoon watersheds often serve to increase the succession rate of lagoons towards their ultimate terrestrial end-state. All of these general processes appear to be currently operating within Chilka Lake.

Primary Producers

Several types of semi-aquatic emergent macrophytes (rooted, macroscopic plants) occur in lagoons, including mangrove-swamp, salt marsh and reed swamp varieties. In tropical lagoons, mangroves usually dominate in the sea-water influenced zone and are absent in waters of less than 10 ppt. salinity. Reed swamp varieties are normally associated with freshwater habitats, but can also withstand brackish or saline conditions. Several species of the sedge, Scirpus, for example, can survive in full-strength sea water (36 ppt.) Varieties of submerged macrophytes in lagoons include eel-and turtle-grasses (Zosteraceae and Hydrocharitaceas) in marine regions and grass-like pondweeds (Potamogetonaceae, e.g. *Potamogeton pectinatus*) in brackish regions. Both the submerged and emergent macrophytes support abundant growths of smaller epiphytic algae. Within lagoon phytoplankton communities, diatoms are important components, as well as dinoflagellates, chlorophytes, cryptophytes and other micro flagellates.

Both phytoplankton (microscopic, water-column algae) and macrophytes are important primary producers in Chilka. The first quantitative study of phytoplankton in Chilka lagoon was undertaken by Devasundaram and Roy (1954). These investigators monitored both phyto— and zooplankton every two weeks at five representatives sampling stations in Chilka Lagoon at Balugaon, Kaluparaghat, Rambha, Satpada and Arkhakuda over a 2-year period.

Highest densities of phytoplankton were enumerated at Satpada and Arkhakuda and most were diatoms of marine origin, including *Asterionella japonica* and *Chaetoceros*. Freshwater forms were prevalent at Balugaon, including the blue-green *Anabaena* sp. and the chlorophyte. Spirogyra sp. Highest phytoplankton densities were encountered in the first half of the year (January-June), coincident with highest incident solar radiation, and peak densities occurred in June.

Plant production rates and species community composition both control the available energy supply for biological consumers. Phytoplankton and primary production were studied adjacent to Rambha between February, 1981 and December 1983 by Panigrahi (1985). Two phytoplankton peaks were enumerated over the annual cycle. The first peak occurred in April-May and was dominated initially by diatoms and later by dinoflagellates while the November peak was dominated by blue green and green algae. Measured chlorophyl concentrations varied between 0.94-3.86 mg/m^3; while estimated primary production rate varied between 6.80 $mgC/m^3/hr.$ to 68.82 $mgC/m^3/hr.$ Chlorophyl concentrations were measured by Mishra et al. (1988) adjacent to Balugaon and ranged between 1.3 mg. in January and 2.46 mg. in November. Primary productivity values varied between 8.13 $mg.C/m^3/hr.$ in January and 41.95 $mg.C/m^3/hr.$ in May.

Past investigations of macrophyte productivity and species composition within Chilka lagoon provide a descriptive benchmark against which present-day macrophyte communities can be compared. Earlier investigators (Devasundaram and Roy 1954; Jhingran and Natarajan 1966) noted the presence of numerous macrophyte species, including *Najas gramineae, Ruppia maritima, Gracilaria* sp. (which provides raw material for agar agar), *Potamogeton pectinatus*. *Najas falcioulata* and *Halophila ovata*. Estimated yield for the latter three species in Chilka lagoon was between 518-565 g/m^2. 77-104 g/m^2 and 34-74 g/m^2, respectively (Jhingran 1991). During the 1960's, the macrophyte flora was dominated by the submerged macrophyte *Potamogeton pectinatus* (approximately 78 per cent of the Chilka Lake macrophyte biomass) and the emergent macrophyte *Scirpus articulatus* (Patnaik 1973A). Additionally, the water hyacinth, *Eichornia crassipes* has recently spread to form solid mats in the northern portion of the lagoon, and is also present in distributary channels close to the lagoon outlet.

Aquatic macrophytes, particularly the pondweed *Potamogeton pectinatus*, have increased in recent decades and infest large areas of the northern sector. ORSAC (1990, 1991) have documented the spread of aquatic vegetation (largely *P. pectinatus*) in Chilka through analysis of LANDSAT satellite images. The following estimates reflect an average macrophyte expansion rate of 15 km^2 per year during the period 1973-1985:

	Estimated Macrophyte Area (km²)
1973	20
1977	60
1982	100
1985	200

Additionally, the biomass of *Potamogeton* measured on an arial basis, has increased by an order of magnitude in recent years and is currently estimated at 4500 g/m^2 wet weight in the central sector during January-February 1991 (ORSAC, unpublished data) compared to about 350 g/m^2 for the same months during 1963-64 (Patnaik 1973A).

A progressive decrease in salinity conditions provides a possible explanation for the rapid macrophyte expansion in the northern portion of Chilka Lake, where freshwater conditions prevail. However, *Potamogeton* and other freshwater macrophytes are tolerant of brackishwater (Twilley and Barko 1990), and an alternative explanation, that the macrophyte expansion in Chilka Lake is related to increase nutrient loading from agricultural run-off, can not be rejected. In fact, a study of the distribution of nutrients in the surface sediments of Chilka Lake suggested a north-south gradient in sediment nutrient concentration (Patnaik 1985), consistent with the latter hypothesis. Measured total nitrogen concentrations in sediments ranged from 23.7 mg/100 mg. to 53.8 mg/100 g. in the northern sector, compared to values of 9 mg./100g. to 25.3 mg./100 g. in the central sector and 14.0 mg/100 g. to 22.0 mg./100 g. in the southern sector. Resolution of whether the macrophyte growth is related to altered salinity concentrations or altered nutrient concentrations in critical for future habitat management within Chilka, particularly for evaluating the feasibility of controlling plant growth through manipulations of lagoon salinity (e.g. via. dredging)

The food chain of Chilka Lake is poorly described (see below). While species descriptions exist for the different trophic levels, the functional relationships between macrophytes, herbivores, detritivores and carnivores are unavailable. Published information on decomposition rates of *P. pectinatus* (Howard-Williams and Davies 1979) suggests that this macrophyte rapidly breaks down to form usable detritus in aquatic food chains. A number of commercial species, including mullets and prawns, consume aquatic plants and detritus directly. There is a high likelihood that the spread of aquatic vegetation has stimulated the production of herbivorous species within Chilka Lake, mediated by increases to their food supply.

Time series records for fish landings do not support the notion that the spread of vegetation has had a deleterious effect on the overall fish production within the lagoon. While it is likely that the expansion of vegetation has affected fish production through alterations in fish community structure, this impact is confounded by simultaneous changes to lagoon salinity conditions and harvesting effects associated with operation of the fishery.

Various species of birds, including ducks and geese, feed directly on macrophytes within the lagoon, or indirectly on insects (118 bird species out of 151 species in total- Dev, 1992) which may be associated with macrophytes. It is, therefore, unlikely that the macrophyte expansion has produced detrimental impacts on the bird fauna; rather, it may have stimulated bird production. From an avian perspective, macrophyte presence within the lagoon in probably an ecological benefit.

Secondary Producers

Both zoo plankton and benthic organisms (secondary producers) provide an important trophic link between the plant and detrial carbon sources produced within Chilka Lake, and the fish species which consume them. Zooplankton populations of lagoons are described by Barnes (1980) as irregular and poorly understood. Typical planktonic forms (e.g. copepods, ostracods) are often replaced in lagoons by small nektonic (free-swimming) organisms (e.g. mysids) and benthic (bottom-dwelling) forms of planktonic larvae and benthic organisms themselves. Within the brackish zone of lagoon, a benthic fauna of mixed ancestry is present with the majority of groups originating in the marine environment, but with a large number of freshwater forms also present (e.g. insect larvae, pulmonate snails, flatworms and leeches). The benthic macrofauna is dominated by three groups: annelids, crustaceans (especially isopods, amphipods and decapods) and molluscs (gastropods and bivalves). Most benthic species are associated with the sediments (infauna and epifauna) or are associated with the macrophytes. Two groups which dominate the nektonic macrofauna of lagoon are prawns and fish. In tropical lagoons, penaeid prawns (e.g. *Penaeus, Metapenaeus*) typically breed in the adjacent sea and enter lagoons as post larvae, and migrate out of coastal lagoons when they are from half to fully grown. A large and diverse fish fauna occurs in lagoons including freshwater, marine and diadromous (migratory between both environments) species. The abundance and seasonal production patterns of both groups can potentially limit fish population size within the lagoons by setting a constraint on the available fish food supply.

Devasundaram and Roy (1954) enumerated zooplankton at five stations in Chilka Lake and provide an initial zooplankton species list. Copepods were numerically predominant amounting to 43 per cent and 70 per cent of their collections from 1950 and 1951, respectively. Other numerically important zooplankters included gastropod veligers, nauplius larvae, bivalve larvae and a diverse group of largely marine species, as well as fish eggs and larvae. Jhingran and Natarajan (1966) report peak densities of zooplankton in Chilka Lake in October-November, dominated by the following species of copepods:

Brackish Copepods	*Marine Copepods*	*Freshwater Copepods*
Pseudodiaptomus annandalei, P. hickmani.	*Isias tropica*	*Diaptomus sp.*
P. binghami, Acartia Ohilkensis	*Labidocera pavo*	*Cyclops sp.*
A. southwelli, A.sewelli Stenhelia inopinata, Laophonte sp.	*Acartia centrura Oithona brevicornis*	*Mesocyclops sp.*

(contd.)

Brackish Copepods	Marine Copepods	Freshwater Copepods
Nitoera spinipes,	*O. nana.*	
Paracalanus crassirortris		
Cyclopina intermedia,	*Paratagestes sphaerius*	
Acartiella major		
A. minor	*Saphirella sp.*	

The latter genera of freshwater copepods were encountered during flood conditions. Patnaik (1973 B) provides the numerical zooplankton and phytoplankton observations collected during his field study (April 1964 to March 1965). A more recent study by Patnaik (1986) confirms that copepods dominate the zooplankton in most of the lagoon except for the southern sector, where Luciferidae and Protozoea are numerically abundant and the northern sector with numerous mysids.

Benthic organisms include plants and animals of restricted mobility that live within or upon the bottom sediments. In addition to their importance in supplying energy to higher tropic levels (e.g. fish), benthic organisms are sensitive to environmental changes, and are frequently used as indicator organisms to monitor the severity and spatial extent of population within aquatic habitats. A diverse group of benthic organisms occur within Chilka Lake including foraminifera, nematodes, polychaetes, copepods, ostracods, isopods, amphipods, gastropods, bivalves, benthic algae and diatoms (Jhingaran and Natarajan 1966). The following annual average benthic biomass estimates were reported by Jhingran and Natarajan (1966).

	Average-Benthic Biomass (g/m^2)
Northern sector	11.1
Central sector	18.3
Southern sector	11.1
Outer channel	13.8

Patnaik (1971) provides a taxonomic breakdown of this benthic biota and indicates a prevalence of molluscs (both gastropods and bivalves) in the benthic community of Chilka. More recently, Mohapatra (1988) analyzed benthic invertebrate distribution, abundance and community structure in Chilka lagoon. A total of 62 texa of invertebrates were identified. Excluding two stations with large Stenothyra populations, the average density in 31 Peterson grab samples was 122 individuals per 0.04 m^2 sample. Based on their faunal similarity, Mohapatra (1988) classified Chilka lagoon benthic invertebrates into three communities.

(i) a "Modiolus undulata community" represented by a group of stations in the Rambha area and associated with vegetation growing on sandy or clayey-sand bottoms with *Nematonereis unicornis, Ancistrosyllis constricta, Clymene annandalei* cirolanid isopods and *Cassidina pulchra* as characterizing species;

(ii) a "Neosolen aquaeducioris community" (characteristic species not yet identified);

(iii) a "Stenothyra blanfordiana community" dominated by this hydrobiid snail and associated with *Clementia vitheleti*, Corophium sp. and tanaids. This community was prevalent in mixed or predominantly silty bottoms in the central and northern sectors.

Fauna of the Lagoon

Chilka lagoon bearing the transition zones between freshwater and marine habitats work as the feeding and breeding grounds for a variety of commercially important fishes, molluscs, prawns, crabs etc.

The fauna of the lagoon can be described as :

(a) Planktonic fauna
(b) Phytal fauna
(c) Bottom fauna
(d) Fish, prawn and crab fauna
(e) Amphibians and reptiles
(f) Avifauna

Planktonic Fauna

The knowledge of planktonic fauna and its eco-biology in assessing the production potential of the lagoon in general and fish resources in particular, is significant.

Chilka lagoon consists of freshwater, brackishwater and marine plankton which appear and disappear depending on the salinity loads. The phyto-plankton in the lake as a whole showed two peak seasons, the principal one being from March to June with dominant species like *Chaetoceros* spp. and *Asterionella japonica*. The second peak is of short duration in November with many forms as *Chaetoceros* spp. and *Bacteriastrum hyalinum*. The zooplankton crop for the whole lake also show two modes closely following the phytoplankton and the copepods for the bulk of zooplankton.

Both high turbidity and sudden dilution by the influx of freshwater is responsible for the low production of plankton during July to October in the lagoon.

The lagoon has a good number of plankton varieties, the abundance of which fluctuate from season to season with average plankton biomass in the range of 0.006 to 0.073 ml/1. From April to August, the lagoon has more of phytoplankton than the zooplankton and from October to November, zooplankton are more than the phytoplankton indicating inverse relationship between phyto-plankton and zooplankton.

The planktonic animals are dominated by the following groups:

1. Copepoda
2. Cladocera
3. Ostracoda
4. Amphipoda
5. Tenaidaceae
6. Cypris larvae

7. Mysidacae
8. Luciferidae
9. Protozoea
10. Alima larvae
11. Medusae
12. Polychaeta
13. Gastropod larvae
14. Bivalve larvae
15. Pteropod
16. Chaetognatha
17. Tunicates
18. Fish eggs and larvae

Phytal Fauna

The phytal biotope forms a separate habitat alongwith the pelagic and benthic biotopes of the aquatic ecosystem. In the weed infested areas the phytal organisms contribute more than the benthic animals towards fish production. Both meio and marco fauna come under this category. The phytal organisms are attached to the weeds like *Potamogeton, pectinatus, Halophyla ovata, Gracilaria lichenoides*. The phytal fauna of the lagoon is represented by the following groups.

1. Foraminifera
2. Nematoda
3. Oligochaeta
4. Polychaeta
5. Ostracoda
6. Copepoda
7. Tenaidacea
8. Isopoda
9. Amphipoda
10. Nauplii
11. Insect larvae
12. Halacaridae
13. Acarina
14. Kinorhyncha
15. Gastropoda
16. Bivalva

Bottom Fauna

The bottom fauna, all those organism being associated with sediment for most part of their life cycle, form a very important source of food for certain groups of fishes. Their distribution and abundance in the lagoon are accounted by the following groups.

1. Foraminifera
2. Nematoda
3. Rotifera
4. Polychaeta

5. Oligochaeta
6. Hirudinea
7. Cladocera
8. Ostracoda
9. Copepoda
10. Cumacea
11. Isopoda
12. Amphipoda
13. Crustacean nauplii
14. Diptera larvae
15. Other insect larvae
16. Arachnida
17. Gastropoda
18. Pelecypoda

The microbenthos from the whole lagoon showed dominance of foraminifer and nematods, whereas among macrobenthos molluscs (83 per cent) are dominant followed by polychaete. The mean biomass from the whole lake is 13.57 gm./sq.m. and the density 2,720 specimens/ sq.m.

In terms of annual mean standing crop (net weight) of bottom fauna for the whole lake is 135.72 kg./ha; the central sector being comparatively richer 182.79 kg./ha. followed by the outer channel 137.96 kg./ha., the northern sector 111.25 kg./ha. and the southern sector 110.88 kg./ha.

Compared to the normally rich natural lake yielding 300 kg./ha. of bottom fauna, Chilka lake appear to be rather poor.

Thienemann (1925) has classified a lake bed producing 1000 animals or less per m^2 as Oligotrophic, one producing above 2000/sq. m. eutrophic and the one producing a number between the above two as mesotrophic Chilka lagoon producing an average number of macroscopic bottom fauna of 2720/ sq. m. can be described as an eutrophic lake.

Molluscs (83 per cent) play a very important role in the food chain of certain fishes like *Gerres setifer, Sparus sarba* as one of the major food items.

Vegetation has a direct bearing on the distribution of many of the benthic forms.

The whole of western shore extending upto about 6 km. in the northern sector and about 1 km. in the southern sector have thick growth of vegetation. Similarly, the eastern shore of southern and central sectors are also covered with vegetation. The rocks and boulders around islands which come in contact with water are covered with luxuriant growth of algae which harbour a number of animal forms. During monsoon many of the submerged weeds die, get uprooted due to the heavy wind and wave action. These dead weeds float on the surface in masses like rafts and are densely populated with amphipods, isopods and molluscs. The outer channel in general is free from thick growth of vegetation.

Fish, Prawn and Crab Fauna

The Chilka lagoon harbours 225 species of fish, prawn and crab and is an important source of fish, prawn and crab in Orissa. Of these, 158 species of fishes and prawn have been identified from the lagoon of which 27 species belong to freshwater and the rest are of marine origin.

Mohanty (1973) reported the occurrence of three more freshwater fishes and five marine species from the lagoon as detailed below.

1.	*Clarias batrachus* (Linn)	:	From Bhusandapur-Kaluparaghat area; local name—Mugura.
2.	*Labeo calbasu* (Ham)	:	From northern sector in the vicinity of river Daya; local name— Kala-bainshi.
3.	*Notopterus chitala* (Ham)	:	Near Kaluparaghat and in the vicinity of Daya river; local name—Chitala.
4.	*Caranx gallus* (Linn)	:	Near lake mouth during winter; local name—Jhanjara.
5.	*Caranx leptolepis* (Cuv.& Val)	:	From outer channel; local name—Kanti.
6.	*Caranx gymnostethoides* (Blkr)	:	From lake mouth, local name—Kanti.
7.	*Equula splendens* (Russel)	:	In the outer channel local name—Karachandi.
8.	*Ichthyscopus incrme* (Swainson)	:	Near lake mouth, local name—Balighumara.

Among crustaceans, *Penaeus monodon, Penaeus indicus, Metapenaeus monoceros, Metapenaeus dobsonii, Metapenaeus affinis, Penaeus semisulcatus* and *Palaemon rudis* are known as commercially important prawns of the lagoon.

The lagoon forms the breeding ground for 44 species of fish besides Gobiids and some mullets. Fishes of commercial importance contributed more than 54 per cent of the average annual landing of the lagoon. More than 50 per cent of the lagoon's commercial species need sea water for completion of their life cycle.

Successful lake-ward or sea-ward migration essentially depend upon the salinity gradient, which decreases from about 34 parts per 1000 to a very low level and a gradual rising and falling salinity gradient exercises directive influence on the migration of Chilka fish either way.

Different species of crab, which occur in fisheries of Chilka lagoon include *Scylla serrata, Neptunus pelagicus, Varuna litterata, Ocypoda sp., Paratelphusa sp.* The mud crab (*Scylla serrata*) is most important species occurs in great number than of other species combined. *Neptunus pelagicus* appears during summer month as a casual visitor, from the sea to the lagoon.

Till now, 158 species of fish and prawn have been identified of which 27 species belong to freshwater and remaining are of brackishwater and marine in origin.

They find abundant food and shelter in the lagoon to grow. The chief immigrants to the lagoon are mullets, prawns, milk fish, sand whiting, sea bass etc. and the endemic forms represented by clupids, polynimids, perches, scianids cichilids etc.

The most important conditions pertinent to the fisheries of the lagoon are favourable environment, food for their growth and access to move to the sea for spawning and the same for the entries of post larvae and juveniles of all varieties into the lagoon.

Important Fish and Prawn Species Commercial to the Chilka

Sl.No.	*Scientific name*	*Common name*	*Local name*
1.	*Mugil cephalus*	Mullet	Khoinga/Kabla
2.	*Liza macrolepis*	Mullet	Dangla
3.	*Eleutheronema tetradactylum*		Sahal
4.	*Hilsa ilisha*	Indianshad	Ilishi
5.	*Nematolosa nasus*		Balangi
6.	*Mystus gulio*	Catfish	Kontia
7.	*Tachysurus arius*	Catfish	Singda
8.	*Osteogeniosus militaris*	Catfish	Sunga
9.	*Plotosus canius*	Catfish	Kanda
10.	*Pseudosciaena coibor*	Jew fish	Borogo
11.	*Lates calcarifer*	Cock-up	Bekti
12.	*Gierres setifer*		Jogli
13.	*Etroplus suratensis*		Kundal
14.	*Sparus sarba*		Dhala khuranti
15.	*Crenidens crenidens*		Horibolia khuranti
16.	*Peneus indicus*	Prawn	Kantal chingudi
17.	*Penaeus monodon*	Prawn	Bagda chingudi
18.	*Metapenaeus monoceros*	Prawn	Sorada chingudi
19.	*Metapenaeus dobsoni*	Prawn	Nika chingudi

Identified and Recorded Fish Fauna from the Chilka Lagoon

Sl.No.	*Scientific name*	*Local name*	*Habitat*
1.	*Lutijanus johnii* (Bloch)	Angarua	Brackish-Marine
2.	*Cynoglossus brevis* Gunth	Ashwa	Marine
3.	*Chatoessus chacunda* (Ham Buch)	Baban Balangi	Brackish
4.	*Eutropiichthys vacha* (Ham.)	Bacha	Fresh
5.	*Eleutheronema tetradactylum* (Shaw)	Baisali	Brackish-Marine
6.	*Lates clacarifer* (Bloch)	Bhekti	Brackish
7.	*Nematolosa nasus* (Bloch)	Balanga	Brackish
8.	*Wallago attu* (Bl & Schn)	Balhia	Fresh
9.	*Ochichthus boro* (Ham. Buch)	Balipoka	Brackish

(contd.)

Sl.No.	Scientific name	Local name	Habitat
10.	*Clossogobius ciuris* (Ham.Buch)	Baligirida	Fresh
11.	*Mugil tade* (Rorsk)	Baranga	Brackish
12.	*Polynemus indicus* (Shaw)	Bhusa sahal	Brackish-Marine
13.	*Seiena albida* (Cuv & val.)	Boraga	Brackish-Marine
14.	*Mugil specleri* (Bleek)	Chanra	Brackish
15.	*Monodactylus argenteus* (Linn.)	Chandi	Brackish
16.	*Colisa lalius* (Ham)	Chandi Kou	Fresh
17.	*Anthoviella tri* (Bleek)	Chanuli Patua	Brackish
18.	*Caranx niurininnis* Day	Charka	Marine
19.	*Liza caeruleo-maculatus* (Lacep)	Chenga Kabla	Brackish
20.	*Aetobatis narinari* (Euphra)	Chili	Marine
21.	*Mugil cunnesius* (Cuv & Val)	Chola	Brackish
22.	*Muraenesox cinerus* (Torsk)	Danti	Brackish
23.	*Liza troschelli* (Bleek)	Dangla	Brackish
24.	*Chrysophrys sarba* (Forsk)	Dhala Khuranti	Brackish
25.	*Ophicephalus punctatus* (Bloch)	Gadisha	Fresh
26.	*Therapon jarbua* (Forsk)	Gahana	Marine
27.	*Tylosurus strongylurus* (V.Hass)	Gania	Brackish
28.	*Mystus vittatus* (Bloch)	Gandia/Gagar	Brackish
29.	*Sphyraene raghava* (Chau)	Gayala	Marine
30.	*Sciena russellii* (Cuv.)	Golra	Marine
31.	*Hilsa ilisha* (Ham)	Ilishi	Brackish
32.	*Gerres setifer* (Ham.Buch)	Jagli	Brackish
33.	*Pangasius pangasius* (Ham)	Jalanga	Fresh
34.	*Lepidocephalicnthys guntea* (Ham.)	Jailari/ Jhimkardi	Brackish
35.	*Sillago sihama* (Forsk.)	Jhudanga	Marine
36.	*Mugil cephalus* (Linx)	Kabla/Khainga	Brackish
37.	*Chrysopnrys datnia* (Ham. Buch)	Khuranti	Brackish
38.	*Tarisseclas kempi* (Chau.)	Patua	Brackish
39.	*Mystus gulio* (Ham.)	Kantia	Brackish
40.	*Caranx carangus* (Br)	Kanti	Marine
41.	*Scatophagus argus* (Bloch)	Kara chandi	Brackish
42.	*Scomber microlepidotus* (Day)	Kanaguda	Marine
43.	*Coius guadrifasiatus* (Sevas)	Katha Khuranti	Brackish
44.	*Thrissoles annandalei* (Chau.)	Kathi patua	Brackish
45.	*Caranx jarra* (Cuv. & Val.)	Katha parei	Marine
46.	*Plotosus canius* (Ham. Buch)	Kaunda	Marine
47.	*Hilsa kanagurta* (Bleek)	Kali pila	Brackish
48.	*Puntius ticto* (Ham.)	Kerandi	Fresh

(contd.)

Sl.No.	Scientific name	Local name	Habitat
49.	*Liza corsula* (Ham.Buch.)	Kekenda Kiakandha Kiakarandha	Fresh
50.	*Chorinemus sancti-petri* (Cuv.& Val.)	Khadisa	Marine
51.	*Anchoviela indica* (V.Hass.)	Kokali	Marine
52.	*Pristipoma argenteum* (Porsk)	Kobarba	Marine
53.	*Anabas testudineus* (Bloch)	Kou	Fresh
54.	*Etroplus suratensis* (Bloch)	Kundala	Brackish
55.	*Carcharinus gangeticus* (Mull & Henle)	Magar	Marine
56.	*Echeneis naucrates* (Linn.)	Magar joka	Marine
57.	*Mugil subviridis* (Cuv & Val.)	Menjia	Brackish
58.	*Cirrhina mrigala* (Ham.)	Mirkali	Fresh
59.	*Rita hastata* (Gunth.)	Musiari	Fresh
60.	*Elops Saurus* (Linn.)	Nahama	Brackish
61.	*Megalops cyprinoides* (Brouss)	Paniakhia	Brackish
62.	*Notopterus notopterus* (Pall.)	Pahli	Fresh
63.	*Gazz minuta* (Bloch)	Polanga	Marine
64.	*Kowala coval* (Cuv.)	Ranji patua	Marine
65.	*Labeo rohita* (Ham.)	Rohi	Fresh
66.	*Hemiramphus gaimardi* (Val.)	Sargara	Brackish
67.	*Ophiocephalus striatus* (Bloch.)	Seula	Fresh
68.	*Chanos Chanos* (Forsk.)	Seba Khainga	Brackish
69.	*Trygon uarnak* (Forsk.)	Shankusha Shankucha Sankara	Marine
70.	*Triacanthus brevirostris* (Temm. & Schl.)	Shukura	Marine
71.	*Tachysurus caelatus* (Val.)	Singda	Brackish
72.	*Liza borneensis* (Bleek)	Soroda	Brackish
73.	*Osteogeneiosus militaris* (Linn.)	Sunga	Marine
74.	*Platycephalus indicus* (Linn.)	Takara	Marine
75.	*Leiognathus eguulus* (Forsk)	Tanki chandi	Marine
76.	*Caranx rottleri* (Bloch.)	Thumburda	Marine
77.	*Macrognathus aculeatus* (Bloch.)	Todi	Fresh
78.	*Catla catla* (Ham.)	Vakura	Fresh
79.	*Epinephelus tauvina* (Forsk)	Vola	Marine
80.	*Penaeus indicus* (White prawn)	Kantal	Marine
81.	*Penaeus monodon* (Triger prawn)	Kalipoi or Bagda	Marine
82.	*Metapenaeus monoceros* (Fabricus)	Chunrdi	Marine

(contd.)

Sl.No.	Scientific name	Local name	Habitat
83.	*M. Offinis* (Milne-Edwards)		Marine
84.	*M. Dobsoni* (Miers)		Marine
85.	*M. Palaemon rudis* (Hellen)		Brackish
86.	*Macro brachium malcomsoni* (Milne Edwards)		Fresh
87.	*Leander styliferus* (Milne-Edwards)		Marine

Biological Outlines of Fishes

Biological outlines of commercial fisheries of the lagoon are:

Mugil cephalus fishery of the lake show fluctuations. Along with *Liza macrolepis* and other varieties of mullets constitute 11.4 to 18.3 per cent of the total catch from Chilka during 1949-50 to 1954-55, the average being 14.25 per cent of the total catch. The percentage composition of mullets out of total catch were from 11.94 to 26.95, the average being 19.07 per cent during 1957-58 to 1965-66 indicate the increase of the group out of total fish population of the lagoon during the period. The group show a decreasing trend among total fish population in the lagoon from 1980-81 onwards till 1990-91, when the group was represented by 10.80 to 19.22 per cent of the total catch, the average percentage during the period being 14.29.

Janos and net fishing are used for capture of the species. The fish favour algal (25 per cent) and detritus (48.5 per cent) as food and attain an average length of 307 mm. in its first year, 512 mm. in second and 662 mm. in third year.

The breeding season of the fish is between September and January with peak breeding month as November. The matured males and females of two and half years age of 238-463 mm (males) and 313-613 mm. (females) migrate in the sea through outer channel of the lagoon for breeding.

From March in certain years and from April onwards juveniles of the species (112 mm. modal size) start lagoon ward migration for recruitment to the fishery and virtually size range upto 499 mm bears the bulk of the fishery. As the recruitment of the species of average eight month old varied greatly from year to year, a good fishery of the species is expected with successful recruitment in the previous year. Poor recruitment of juveniles of the species led to poor or nearly poor fishery in succeeding years.

The mean length in the commercial catch during 1957-65 was 318 mm., the maximum observed length being 712 mm. During 1987-91 the mean length of the species in commercial catch has been observed as 304 mm.

Mass destruction of brood fish at lake mouth during October-December period by Kadijal and the predation of developing eggs of mullets by *Ambassis gymnocephalus*, which abound in inshore water during their breeding period are the main reasons for recruitment failure of the species to the lagoon.

Liza macrolepis forms 1.3 per cent of net catches and 12.48 per cent of Jano catches. Rod and line are extensively used with green algae, chiefly *Spirogyra* as bait in winter months in the main lake and outer channel. Bulk of catches are made in July-September period with size upto 610 mm.

The fish feed on algae (45.6 per cent) and detritus and reaches an average length of 224 mm. in its first year, 348 mm in second year and 470 mm. in third year. Males at 315 mm. and females at 415 mm. attain maturity and move out of the lagoon from November to January with peak breeding month as January. Developing eggs of this species are observed in large numbers near lake/mouth in November-December indicating inshore breeding of the species close to lagoon inlet area. Huge waves of fry (8-30 mm.) of this species have been noted to move into outer channel in the month of January. Migration (Lake-ward) of juveniles is more after sun-set and in full moon and new moon nights. *Eleutheronema tetradactylum* breeds both in the sea and the lagoon and performs sea-lagoon and vice versa migrations. Sea-ward breeding migration takes place between July to September. Two broods participate in the fishery of the lagoon, one of which is the most dominant. Some breeding of the species take place during February to June in the northern sector of the lagoon opposite to Daya river mouth off Badakudi village in Tua nala. The major spawning of the species takes place in the sea.

The species feed on prawns (49.5 per cent), fishes (26.1 per cent), mysids (9.8 per cent), stromatopods (2.9 per cent), isopods (2.1 per cent), amphipods (1.3 per cent) and miscellaneous matter and grow to an average length of 189 mm. in first year, 369 mm. in second year, 537 mm in third year and 632 mm. in fourth year.

Recruitment to the fishery takes place twice a year, first during February-April in the size range 62-112mm. and again in August-September at 112 mm. Of the two recruitments noted in Chilka, that of the period February-April is much more dominant and it is likely that the recruits are immigrants from the sea.

The size groups 50-249 mm and 250-449 mm bear the bulk of the fishery. By number the first group is by far the most dominant. Larger size groups measuring 575 mm and above and to a large extent of 450-474 mm. size groups make their appearances in the catches seasonally during the period of November to May and more especially in January to May. The maximum size noted in the commercial catch was 1000 mm.

The percentage of this species in total catch of the lagoon during 1949-50 to 1954-55 accounted for 2.7 per cent at the lowest and 6.7 per cent at the highest with an average of being 4.42 per cent. Its population in total catch increased during 1957-58 to 1965-66, when its occurrence in the total catch was registered as 3.80 per cent as minimum and 10.83 per cent as maximum with an average of 6.44 per cent during the period. Thereafter, the contribution of the species in total catch showed a decrease in 1980-81 with marginal variations till 1990-91, when the lowest 3.07 per cent and a highest of 5.27 per cent out of the total catch was noticed with an average of 4.42 per cent.

Hilsa Ilisha — fishery of the lagoon showed considerable fluctuations, and the landings are generally prominent in winter/spring as well as monsoon months though they occur in the lagoon throughout the year.

Breeding of the fish appears to take place between August and October and the fish ascends the river Daya for breeding through Tua-nala and a separate stock is maintained for the lagoon. There are two main waves of migration of fish into the lagoon, one at the close of winter and the other at the commencement of monsoons.

The fish is captured by nets and 245-480 mm., 110-395 mm., 290-330 mm.,

125-500 mm., and 205-520 mm., size groups dominantly occur in bhidajal, menjia jal, hilsa jal, patua jal and in patua bhida jal respectively.

The overall mean in commercial catches was estimated at 259 mm.

The recruitment of the species to the fishery is highly variable and takes place in December-January at modal sizes 50 to 149 mm.

The fish feed on 31.7 per cent copepoda, 32.0 per cent organic detritus, 17.5 per cent prosobranch and lamellibranch larvae, 9.1 per cent mysids in addition to diatoms.

The shallow sandy areas on the eastern side of the southern sector of the lagoon provide favourable breeding grounds to *Nematolosa nasus* as eggs, larvae and juvenile stages of the fish have been observed from the various parts of the lagoon.

The fish move into the lagoon from the sea during January-June at size ranged from 100-355 mm., dominant sizes, being 135-150 mm. and 195-255 mm. This lagoon ward immigration of the species also coincides with its peak catches during the period.

The fish prefers rotifers, algae, sand and mud as diet. Decayed organic matter (63.5 per cent) copepods crustaceans etc. (11.9 per cent), diatoms (4.5 per cent) and algae (3.5 per cent) form the main bulk of their food and grow to an average length of 67 mm. in first year, 154 mm in second year and 232 mm., in third year.

The recruitment of the species to the fishery takes place at modal sizes of 37 to 67 mm. during March to May and the success of the fishery is chiefly contributed by the second year age-group.

Males at 124 mm. and females at 116 mm. mean size attains maturity and the annual mean length in commercial catches vary between 135-179 mm.

Mystus gulio breeds during June to November with prolonged breeding season. The fish feed on amphipoda (30.8 per cent), prawns (13.1 per cent), algae (12.6 per cent), detritus and decaying higher plants (5.3 per cent), fish (4.8 per cent), mysids (3.8 per cent), gastropods (13.6 per cent), isopods (3 per cent), insects (2.3 per cent), copepods (1.7 per cent), lamellibranchs (1.3 per cent) and miscellaneous items (32.3 per cent) and grow to an average length of 78 mm. in first year, 115 mm. in second year, 145 mm. in third year, 175 mm. in fourth year.

Recruitment of the species to the fishery takes place in May to September at modal sizes of 75 to 85 mm. The impact of fishery is pronouncedly felt on four-year age group and the overall mean length in commercial fishery is 131 mm. The maximum size observed in the commercial catch was 260 mm.

The fish at 82 mm. attain maturity, much below the overall mean length in commercial fishery. The lagoon appears to favour the reproduction, growth and abundance of the species.

The species alongwith other catfishes showed their dominance in the total catch of the lagoon by 3.2 to 10.1 per cent during 1949-50 to 1954-55 with an average of 6.5 per cent of the total fish production.

Their population in the lagoon increased year by year as noticed from their percentage in catch from 10.1 to 15.85 with an annual average of 12.40 per cent during the year 1957-58 to 1965-66.

The abundance of the species dropped down to 6.24 per cent in 1980-81 and 1981-82 and again registered an increase from 1982-83 by 11.81 per cent with marginal fluctuation between the years, till 1988-89 from where a steady increase of their population in the catch (13.67 per cent) was noticed year by year until the highest percentage in their catch was observed in 1990-91, being 17.89 per cent. The average percentage composition of this species during these years were calculated to be 11.55.

Recruitment of *Lates calcarifer* to the fishery of the lagoon takes place in February at 162 mm. size (modal value) and again in July-August at modal sizes of 112 to 162 mm. and is highly variable.

Maturing and matured specimens of the fish have been observed during April-July in the lagoon. The spawning of the species is believed to take place in the inshore waters of the sea, with peak breeding in June and July.

Lates calcarifer feed on fish (51.6 per cent), prawns (39.3 per cent), stomatopods (7 per cent) and miscellaneous matters including weeds (2.5 per cent).

The February recruit attains average length of 287 mm. in first year, 492 mm. in second year, 687mm. in third year and 797 mm. in fourth year. The July-August recruits reach an average length of 317 mm. in one year and 517 mm. in second year. Sizes larger than 837 mm. occur with little abundance. The annual mean length in commercial catches varied in the range of 234-562 mm. and the minimum size at maturity in 425 mm.

Peak catches of *Gerres setifer* from the lagoon are landed during the period November-May.

Evidences show that the species breed inside the lake, especially in the southern sector of the lagoon in May-August. The lagoon-ward migration occur during November-May.

The fish feed on crustaceans (42 per cent), molluscs (31 per cent), algae (6 per cent), decayed organic matter (13 per cent) and miscellaneous matters (8 per cent) and reaches an average length of 50 mm. in first year, 109 mm. in second year and 172 mm. in third year.

The common size at recruitment is 52 mm. (modal value) though smaller sizes (37 mm) also occur sparingly. The recruitment occurs in March, May, August, November and December, indicating extended breeding.

The mean length in commercial catches varied in the range of 84-145 mm, the maximum length in the commercial catch being 232 mm. The fish attain maturity at 73 mm. in case of males and 86 mm. in case of females.

Perches constituted 7.2 to 13.0 per cent of the total catch of the lagoon during 1949-50 to 1954-56 with an average of 9.3 per cent during the period. During 1957-58 to 1965-66, the group registered higher percentage of catch to 11.03 per cent (average) with highest of 29.19 per cent in 1964-65.

In the years between 1980-81 to 1990-91 the group occupied 6.95 per cent to 15.64 per cent out of the total catch of the lagoon with higher average of 13.07 per cent from the previous years.

Pseudosciaena coiber form a good fishery in the lagoon even before commencement of monsoons i.e. February-June. In certain years the fishery is good in flood and post flood periods. The fish breeds in the northern sector of the lagoon in summer months (April-July) in addition to its sea breeding around March.

The species feed on prawns (35 per cent), fish (20.7 per cent), amphipods (15.1 per cent), isopods (8.2 per cent), stomatopods (6.1 per cent), detritus (4.1 per cent), higher plant matters (2.5 per cent), algae (2.2 per cent) and miscellaneous matter (4 per cent) and grow to an average length of 235 mm. in first year, 392 mm. in second year, 582 mm. in third year and 697 mm. in fourth year.

There are two recruits of the species to the fishery of the lagoon, the most dominant, in July-September at 87 to 112 mm. modal size and the weaker one in February at modal size of 112 mm. The July-September recruits are most important from fisheries point of view.

The mean length in commercial fishery varied from 204-363 mm. with maximum recorded size of 800 mm. The minimum size at maturity was 176 mm. in the case of males, and 213 mm. in females.

During 1949-50 to 1954-55, the scianids comprising of the species occurred as 2.9-6.1 per cent of the total catch of the lagoon with a mean percentage of 4.38 during the period. A slight increase in mean percentage (5.90) of this group in total catch has been registered in years 1957-58 to 1965-66, varying 2.85 to 9.41 per cent annually out of the total catch.

Etroplus suratennis formed only an average yield of 4 tonnes during 1948-52 as against the average annual yield of 108 tonnes during 1961-65. In 1990-91, the annual yield of the species rose further to 339.09 tonnes, constituting 7.94 per cent of the total catch.

The fish feed on weeds (48.03 per cent), algae (11.93 per cent), detritus (34.1 per cent), gastropods (3.6 per cent) and other miscellaneous matters comprising bivalves, insects, mysids, sand and mud. Among aquatic plants, *Potamogeton* and algal *Spirogyra* and *Lyngbia* are the chief constituents. They grow to an average length of 105 mm. in first year, 175 mm. in second year and 245 mm. in third year.

Maturing and matured specimens occur nearly throughout the year. The lagoon especially, its western shore, provides favourable breeding ground and breeding takes place in two peaks, one during December-February and the other in April-May.

The recruitment of the species to the fishery takes place in May-June and in October-December at modal size ranges of 35 to 55 mm. and 45 to 55 mm. respectively.

The mean length of the fish in commercial catch is represented in the range of 111-165 mm. The maximum size observed in the commercial catch was 335 mm. The fish attains maturity at a minimum size of 105 mm.

Chilka Lagoon as Feeding and Breeding Ground of Fishes

Each species of fishes though show a specific feeding habit, yet had detritus as one of the component of food in most of the cases.

Similar to any ecosystem, primary producers covert the solar energy into utilizable chemical energy. In Chilka primary producers are mainly comprised of higher aquatic plants, benthic flora and phytoplanktons. Herbivores, the primary consumers use stored up energy directly. Decomposition of flora and fauna leads to formation of organic detritus. Besides the normal food chain, energy flow from one trophic level to another. Another by pass chain formed from primary producers to detritus to consumer of all normal trophic levels, which prove that whenever an ecosystem offers a particular forms of food materials in excess, the constituent consumers try to use it as a part of their food. Fry and fingerlings of fishes with their specific food accept the detritus as a part, when they grow.

Feeding habits of Chilka fish can be grouped as follows:

(a) *Detritus feeder* Some of fish having main food habit of dead and decaying organic matters of plants and animal origin. *Nematolosa nasus Mugil cephalus* (48.5 per cent detritus, 25 per cent algae).

(b) Plankton feeder *Hilsa ilisha* mainly feed on planktons, 39.4 per cent copepods, 12.2 per cent mysids, and 38.2 per cent detritus.

(c) *Herbivores* *Liza macrolepis*, predominantly an algae feeder (45.6 per cent). The food of *Rhabdosargus sarba* constituted by 31 per cent algae with other components like gastropods (20 per cent) and crustaceans (10 per cent). Higher plant matter contribute the major portion (48.6 per cent) of food of *Etroplus suratensis*.

(d) *Carnivores* Both fish and prawn constitute the main food of this group of fish e.g. *Eleuthoronena tetradactylum, Lates calcarifer, Pseudosciaena coiber*.

(e) *Omnivores* All types of food materials are taken by this group representing *Mystus gulio* and *Tachysurus arius*. As gut contents of prawns i.e. *Penaeus monodon* and *P. indicus* mainly comprises detritus due to presence of other plants and animal matters, they are classified under omnivores.

The lagoon provides vital protection and support for many commercial and forage fish during one or more stages of their life cycles forming a nursery ground of about thirty species in addition to the immigrants from the adjoining Bay of Bengal, serving as brackish water fish seed resources for the State. The outer channel system harbours 24 species, the richest composition, and serves as entry path for the fry of prawn, perches and mullets.

Barkudi-Bhusandapur zone, second in species composition, provide breeding and nursery ground for well over 16 species range first, in larval abundance. Rambha zone, Balugaon-Nalabana-Kathapenthaghat zone and Bardihi-Magarmukh zone harbours 15 species and 10 species respectively giving poor abundance in larval numbers than the other sectors. This is correlated by the richness of planktons in northern and outer channel sectors than Rambha zone, Balugaon-Nalabana-Kathapenthaghat zone and Baradihi-Magarmukha zone.

Spawning season lasts for about 4 months for the group like *H.ilisha, K. coval, N. nodus, R. corsula, P. arsius, B. orientalis* and *M. armatus*. Extending spawning season lasting for about six months, have been noticed in fish e.g. *E. saurus, H. kelee, E. tetradactylum. Therapon* spp. *E. sihama, Leiognathus* spp. *Gerres* spp. *Sciaena* spp. *S. argus, Cynoglossus* spp. & *T. brevirostris*.

Table 3.1

Species	*Family*	*Common Name*	*Local Name*	*Average Total Landings 1957-1965 (MT)*	*Average Length (mm)*			*Spawning Location*	*Breeding Season*	*Time of Recruitment to Fishery*	*Modal Recruit-ment Size (mm)*	*Feeding Habits*
					Age I	*Age II*	*Age III*					
Mugil cephalus	Mugilidae	grey mullet	kabla koinga	478	307	512	662	sea	September-January	April onwards	112	herbivorous detrivorous
Liza macrolepis	Mugilidae	mullet	dangla	114	224	348	470	sea	November-January	July-September	110	herbivorous detrivorous
Eleutheronema tetradactylum	Polynemidae	Indian salmon	sahal	238	203	383		sea and lake	January-June	Feb.-April Aug.-Sept.	62-112 112	carnivorous
Hilsa ilisha	Clupeidae	hilsa	ilish	142	217	367		Daya River	August-October	December-January	50-149	plantivorous
Nematalosa nasus	Clupeidae	sardine	balang	141	67	154	232	sea and lake	Febuary-July	March-May	37-67	omnivorous
Mystus gulio	Bagridae	catfish	kontia	311	78	115	145	sea	June-November	May-September	75-85	omnivorous carnivorous
Lates calcarifer	Centropomidae	cock-up	bekti	215	287	492		sea	April-July	Febuary July-Aug.	112 112-162	carnivorous piscivorous
Gerres setifer	Gerreidae	perch	jagili	45	50	109	172	sea and lake	May-August	March-December	52	omnivorous
Pseudosciaena coibor	Sciaenidae	jewfish	borago	199	235	392	582	sea and lake	April-July	July-Sept. February	87-112 112	omnivorous
Etroplus suratensis	Cichlidae	pearl spot	kundal chakkl	107	105	175	245	lake	All year	May-June Oct.-Dec.	35-55 45-55	herbivorous detrivorous

In fish like, *Thrissocles* spp. *Anchoviella* spp. *S. strongylurus, H. gaimardi, I. carce, H. brachyrhynchus, P. panchax, Ambassis* spp. and gobids, the breeding is a continuous process with intermittent slack period.

Seven species exhibit spawning peaks in pre-monsoon season (Feb-May) though the beginning of pre-spawning activities start in many fish. They are *Thrissocles* spp., *Anchoviella* spp., *K. coval, H. gaimardi, S. strongylura, E. tetradactylum* and *Ambassis* spp.

Twenty three species of the lagoon representing *H. ilisha, H. kelee, N. nasus, S. sihma* and *T. breverostris* show peak spawning in monsoon months (June-September).

The spawning peaks of 12 species of Chilka represented by *M. cephalus, L.macrolepis*, other mullets, gobids, besides *H. ilisha, H. kelee Anchoviella* spp. *S. strongylurus* and *H. gaimardi* are noticed in post monsoon and winter months.

Life history features of dominant commerical fish of Chilka lagoon has been compiled as Table 3.1.

Amphibians and Reptile Fauna

Amphibians and reptiles are usually found in various islands of Chilka. Mainly snakes including sea snakes, crocodiles and lizards belong to reptile group of the fauna.

Rana tigrina, Rana hreviceps, Rhacophorum maculatus and *Microhyla ovata* are the important species of amphibians of the Chilka lagoon.

Chelone imbricata, Chelone mydas, Hydrphis spp. *Hydrphis caronatus, Hemidactylus brooki, Hemidactylus feratus, Calotes* spp.*Mabuya* spp., *Varanus bengalensis* and *Varanus salvator* are the important species belong to reptiles of Chilka lagoon.

Avifauna

The Chilka lagoon hosts over 150 species of birds in the peak winter migratory season. Among them 32 per cent are aquatic, 22 per cent waders and about 46 per cent are terrestrial birds. 13 species of birds of prey belong to terrestrial category.

The birds of Chilka lagoon can be divided mainly into two categories (a) Resident birds, (b) Migratory birds. Spoon birds, Darters, white ibis, Cormorants, Herons and Egrets are the important groups among resident birds. Different species of Flamingoes, gull, tern, plovers, snipe, sand piper, pellican, shank, goose, grey lay, teal, pochard, shovellers, dabchicks, spot bills, pin tails, duck, gardwall, malland, pigeon bittern, stilt, water hen, godait and coots are the important groups among migratory birds of the lagoon.

4

Flora of the Lagoon

Phytoplankton

The microflora of Chilka lagoon plays an important role in the bio-mass production of the lagoon. The phytoplankton in the lake consisted of freshwater, brackishwater and marine forms appear and disappear depending on the salinity conditions of the lagoon. They constituted mainly by diatoms, blue-green algae, green algae and dinoflagellates. As a whole they show two peak seasons, the principal one being from March to June (Summer) with little fluctuations. The dominant forms are *Chaetoceros* spp. and *Asterionella japonica*. The second peak season of short duration in November (winter) is due to rapid vegetative multiplication consequent on the rise of salinity and richness of nutrients in the water after monsoon.

Macrovegetation

Plays an important role in the lagoon eco-system. Very few species of macro-vegetation occur in the lagoon and can be broadly divided into two categories

(a) Temporary vegetation
(b) Permanent vegetation

The plants like *Eichhornia crassipes, Pistia stratiotes, Azolla pinnata, Monochoria veginalis* are considered temporary vegetation as they are found during a particular period of the year. They enter Barakudi-Bhusandapur zone of the lagoon during monsoon alongwith flood water and remain there till March. Similarly plants, like *Hydrilla verticillata, Vallisneria spiralis, Nympae* spp., *Utricularia* spp. *Polygonum* spp., *Ipomea* spp., *Jussicea* spp. *Ceratophyllum* spp. *Marshelia* spp. and Duck weeds are observed to grow on the margins of Barakudi-Bhusandapur zone during November to March.

The permanent vegetation includes only five species e.g. *Potamogeton pectinatus, Nagas foveolata Halophila ovata, Halophila beccarii* and *Ruppia maritima*. Of these first three species are widely distributed and are of quantitative importance. In addition to this, marginal weeds like *Eleocharis plantaginia, Scirpus littoralis, Cyperus articulatus* and other grasses remain in the margin of Barakudi-Bhusandapur zone throughout the year.

The weeds listed under temporary vegetation remain restricted to Barakudi-Bhusandapur zone, where the water becomes almost fresh after entry of flood water. As the salinity in this zone gradually rises to 1.9 to 2.4 ppt. by March, the plants are unable to tolerate and die.

Permanent vegetations remain all the year round with seasonal fluctuation in density.

Potamogeton pectinatus constitute 78.2 per cent of the total weed mass of the lagoon. The plant grows luxuriantly from the silty northern to sandy southern end and tolerate wide salinity fluctuations of 0.26-15.00 ppt. Comparatively denser growth are observed on the western shore of the lagoon and the highest density of 3870 gm/sq.m., recorded in Balugaon-Kalijai and Rambha zone. Maximum density reaches in February.

Najas foveolata constitute 14.2 per cent of total weed mass in the lagoon. It grows in patches mixed with *P. pectinatus*. It is distributed in both eastern and western shore of Rambha zone and Baradihi-Magarmukha zone, Balugaon-Nalabana-Kathapenthaghat zone and only on the western shore of Barakudi-Bhusandapur zone. The period of dense growth is during June-August in Rambha zone, April-May in Baradihi-Magarmukh zone and Balugaon-Nalabana-Kathapenthaghat zone and July-September in Barakudi-Bhusandapur zone. The highest density of 3270 gm/sq.m. recorded near Bhusandapur area.

Halophila ovata constitute, on an average 7-12 per cent of the total weed mass and is widely distributed all along the eastern shore of Baradihi-Magarmukh zone, Balugaon Nalbana-Kathapenthaghat, Rambha zone and some parts of the western shore of the Rambha, Balugaon-Kalijai zone. The weed is completely absent in Barakudi-Bhusandapur area. It grows on sandy substratum at less salinity fluctuations. Higher density is observed during April to July in Baradihi-Magarmukh zone, Balugaon-Nalabana-Kathapenthaghat zone and November to June in Rambha zone. The maximum density of 1362 gms/sq.m. recorded near Parikud area.

Halophila beccarii constitute 0.24 per cent of the total weed mass and distributed on the eastern shore of Barakudi-Bhusandapur zone, Baradihi-Magarmukh zone and Balugaon-Rambha-Nalabana-Kathapenthaghat zone. It grows in shallow areas in patches during November-May. Maximum density of 232 gms./sq.m. was recorded in Baradihi-Magarmukh zone and Balugaon-Nalabana-Kathapenthaghat zone.

Ruppia maritima is distributed in the eastern parts of Rambha zone and Balugaon-Nalabana-Kathapenthaghat zone constituting only 0.06 per cent of the total weed mass of the lagoon. Maximum density was recorded as 54 gms./sq.m. from Rambha zone.

Grasses occupies mainly the northern zone of mouth area during November-December. Around each island in Chilka other small grasses are noticed.

Among algal form, the most important one, *Gracileria* spp. grows on sandy bottom, rocks and boulders in Balugaon-Nalabana-Kathapenthaghat, Baradihi-Magarmukh zone and Badakuda, Sanakuda, Kalijai island and inner parts of the outer channel between Gurbari and Alupatna.

Maximum density of 898 gms./sq.m. was recorded between Nalabana and Nuapada area. The other associated alga like *Enteromorpha* spp. have been observed on the leaves and stem of *P. pectinatus* and *N. foveolata* in Baradihi-Magarmukh, Balugaon-Nalabana-Kathapenthaghat and Rambha-Balugaon-Kalijai zone of the lake.

Algal forms like *Spirogyra* spp., *Polysiphonia* spp., *Lyngbya* spp., *Pithophora* spp. and *Cladophora* spp. have been recorded from the lagoon.

Isopods, Amphipods, eggs of molluscs remain associated with aquatic weeds. The insect larvae are predominant in weedy areas. Some fishes remain in densly weed infested areas because of their nest building habit. Eggs of many fishes also remain associated with weeds.

Distribution of Weeds

The distribution of weeds in Chilka varies from season to season, mainly pre-monsoon, monsoon and post monsoon periods. Maximum dense weed growth is found in post monsoon period and the lowest in monsoon period. The distribution of weeds also varies in different zone.

The northern zone covers more area of the lagoon. Due to the differences in the salinity different types of aquatic plants are seen in this zone. On account of distribution of aquatic plants, this region is also divided into three sub-divisions. (a) The area from Mangalajodi-Haridas to northern fringes of the lagoon, (b) Kalupadaghat-Barakudi to Baradi-Panaspada, and (c) Baradi-Panaspada to Kumandol-Tua. As this zone is the shallow region of the lagoon, water depth varied from 0.51 to 1.6 m. The salinity ranged from 0 to 5 ppt. The salinity of Managalajodi-Haridas area was very low ranging from 0 to 0.2 ppt. This area is very shallow compared to the other area of this zone. Most of the freshwater plants are available in this area. It receives more silts in rainy season. The sediment load is more here with rich in nutrients. Submerged, floating and emergent freshwater hydrophytes grow luxuriantly in this region. The submerged plants of this region are *Hydrilla verticillata, Vallineria spiralis, Ceratophyllum demersum, Utricularia spp., Mariophyllum spp. Potamogeon crispus*. These plants found in the deeper region of this area. Floating tracheophytes includes *Eichhornia crassipes, Potamogeton purburacfolia Pistia stratiotes, Nymphoides indicus, N. cristata Nymphea glabra, N. alba, Ipomea aquatica, Jussicea* sp., *Commelina bengalnesis*. Besides these tracheophytes, two floating pteriodophytes such as *Azolla pinnata, Marselia quadriatifolia* are also found. The emergent weeds included *Schoenoplectus littoralis, Phragmites karka, Eleocharis dulce, Macrostachya weight, Fimbristylis spp., Cyperus spp., Typha angustata* and *Panicum repens*. These are found from the fringes of the lagoon to some depth of water. In summer they are found within few centimeter of depth or without water. Some amphibious grasses are also seen in front of Badurabar, Manakgaon, Dochhia villages. *Phragmites karka*, which is locally called as "Nala" or "Bagha Nala", once believed to be found abundantly in Nalaban area, now luxuriantly grows in this region. It is associated with *Schoenoplectus littoralis* and *Typha angustata*. This *Pharagmites* found in patches extending from an area of 20 to 500 sq. meters surrounded by *Scheonoplectus*. This patches were observed at Mangalajodi and Dochhia area. Like many eutrophic lakes in the world, in Chilka, the marginal reed in association of Phragmites provides striking examples of the detailed ecological effect of exposure. *Schoenoplectus* ordinarily grows in 0.92 to 1.5 meters of water usually on the lagoonward side, thus giving rise to some zonation. Here in a semicircular zone of one km. to five kms. width *S.littoralis* are found in Kalupadaghat area to Barakudi area in softer sediments. Since the terete leafless stems of

Schoenoplectus are better able to withstand wind and wave action than is *Phragmites*, this zonation when it occurs, is adoptive. *Typha angustata* is certainly more easily injured by wind or wave than *P. Karka* and is usually found only in well protected places. Wind and waves have the greater role in shallow eutrophic lakes. Since wave action is restricted in northern zone due to heavy submerged weed, the wind has greater role in this region. Due to this reason *T. angustata* is not seen in deeper region of Kalupadaghat Mangalajodi and Bhusandapur area. It is found abundantly in Dochhia area of the lagoon well protected with *P. Karka*. It is found in patches at Kalupadaghat area protected by *P. Karka* and *S. littoralis*.

The existence of emergent reeds provides an environment in which the surface of the water is much less disturbed by wind than in open area and which, is thus, suitable for population of floating plants. These floating plants get shelter and complete their life-cycles with others' associations.

In the zone of floating-leaved vegetation there is a good deal of evidence of the regular effect of disturbance, though it is not easy to separate the wind and wave factors from each other. In case of *Nymphea alba* and *Nymphoides indicus*, *N. alba* is more adapted to wind disturbance than *N.indicus* due to its bigger size of rhizome.

In Chilka with exposed shores, there is erosion, below which there is a neutral zone, where erosion is balanced by sedimentation, whereas further away in deeper water sedimentation prevails.

This zone is very sensitive zone in Chilka because all these plants prevent sediments to move further. Specifically all the submerged hydrophytes act as seive to the flood water. In the post-rainy season, when flood waters enter the lagoon with heavy silt and clay these plants obstruct their movement.

So these clay and silt particles settle down there forming a sediment layer on these submerged macrophytes. In course of time these deposited layers give rise to peats and mucks. Consequently, after some year giving rise to some landforms. The major river which enter the lagoon in the northern fringe have narrow inlets and shallow basins. In the time of heavy rains the whole of the northern fringe becomes inundated with flood water and the water current in the region is very slow compared with the upper region of the rivers. From the limited high current zone silts move to the lagoon and settle down in the area, where current action is very slow.

Kalupadaghat-Barakudi to Baradi-Panaspada sub-zone

In this sub-zone, the emergent reeds are found in patches. They are not compact like other sub-zones. A dense area of *S. littorolis* is observed in the shore of Kalupadaghat. This plant is found upto Soran area in the western side and Barakudi to Panaspada area in the eastern side in small patches. They are not found in the central region. *P. karka* are seen in few patches at Kalupadaghat area only. A single patch of *T. angustata* is observed near Kalupadaghat associated with *P. karka* during macrophyte survey work.

The floating weeds like *Nymphea glabra*, *N. alba*, *Nymphoides indicus*, *N.cristata* *P. purpuraefolia*, *Pistia stratiotes*, *Azolla pinnata* are found in association with emergent weeds. In the central region, where the reeds are completely absent, these floating plants are not found. This might be due to wind and wave action on these plants. *Eichhornia crassipes* are found in the shallower region near the shoreline from

Kalupadaghat to Panchupatia area, Haridas to Barakudi area and Daya, Bhargavi, Nuna confluence area in compactness.

They are not seen Soran, Baredi and Panaspada area of this sub-zone. They grow luxuriantly in shallow zone with in softer sediment bed. In the rainy season this floating weeds come upto cental zone and as soon as the salinity increases it dies. The submerged plants recorded in this sub-zone are *Potamogeton crispus, P. pectinatus, Najas major, N. minor, N. faveolata, H. verticillata* and *V. spiralis*. Except *P. pectinatus* all are found in the shallower region and in less saline area of western side of the lagoon. In the deeper region, where salinity is more *P. pectinatus* is found alone in patches in rainy season and compactly in winter and summer seasons. In Baredi area *P. pectinatus* associated with *N. minor* and *N. major* in deeper region and in shallow region near to shore *Najas* in association with *N. verticillata* have been observed. In the eastern region *Najas* is associated with *P. pectinatus* in shallow region. Sometimes *Hydrilla* is found in small pockets in the eastern region. *Najas* and *Hydrilla* locally known as "Chingudia Dal" are growing in the shallower region on softer sediment where high solar radiation with slow wave and current are present. They are associated with few plants of *Najas* and *P. purpuraefolia*.

Baradi-Panaspada to Kumandal-Tua sub-zone

This is the deeper region of the northern zone. The water depth ranges from 0.5-1.6 meters. The salinity is comparatively high in this region. No emergent or floating aquatic plants are found in this region except some small patches of *S.littoralis* in the shallow fringes of Tua area of eastern side of the lagoon. *Potamogeton pectinatus* locally called as "Charidal" or "Kankada dal" found abundantly in this region. In the post monsoon period they come upto surface of the water and forming the mat in winter and summer seasons. A maximum sizé of 2.6 meter long is recorded during survey. In months of May and June when Chilka experienced the heavy southern wind these plant are detached from the deeper region due to high wave action and float towards the shore-line in rainy season—causing a bad odour; while it decomposes. *P. pectinatus* associated with *Najas* are also found in Bhasara mundia area and in the pockets of Kumandal area.

Central zone

This zone starts from behind of Bhasara hill to Jatia hill in western side to Tua to Gheemundaghat of Parikud in the eastern side. The water depth of this zone ranges from 1.7 to 2.5 meter. The salinity ranges from 4.0 to 9.0 ppt. As it is the deeper with high salinity, compared to the northern zone, most of the aquatic vegetation is submerged type. Wind and wave play great role in this region devoiding floating and emergent plants. However, a thin patch of *S. littoralis* is found in the fringes of Bankiapalli hills to Barakul shoreline. It is also found in a few thin scattered patches near Nalban. *P. pectinatus* associated with *Hydrilla* and *Najas* are found in shallower region of Balugaon. In the deeper region, that is, Nalban area, Kathpentha area, Nuapara and Magarmukh area *P. pectinatus* is available in association with *Najas*. Patches of *S. littoralis* are also found near the Magarmukh and Gerasar area of Nuapara. Some free floating plants like *E. crassipes* are found near the inlets of Kansari and Salia rivers. *Hàlophila oveta*, which is locally called as "Kamari Dal" found in the sandy beds near Nalban area and Kathapenth area, where salinity is more.

Nalban, which comes under central zone specifically known for the migratory birds congration. Locally Nalaban means "forests of *Phragmites*". But now, there is no phragmites in this area. Only some dense grasses such as *Panicum repens* and *P.welence* found abundantly in Nalban. Few patches of *S. littoralis* are also found as emergent plants. Presence of *S.littoralis* in softer sediments and exposure in the summer season indicates, possible existence of *P. karka* in the past. They might have disappeared due to biotic interference such as grazing of buffaloes and cutting them by local people for fuel purposes. Wind might have also played a great role for their disappearance.

Southern Zone

It is started from Jatianasi-Kalijai-Parikud line to the end of the lagoon in south. This area is the deepest zone in Chilka. The water depth varied from 1.65 to 3.4 m. The salinity ranged from 7.0 to 15.0 ppt. Most of the hydrophytes found in the zone are submerged ones. Marginal reeds like *S.littoralis* was found in some places in Rambha area. *P. pectinatus* is the dominant plant available in this zone. *Ruppia mertima H.oveta, H.becari* are available in the sandy bottoms of Rambha, Taltalanasi area and eastern side of this zone. *P. pectinatus* associated with *Najas* are found in the shallow regions. In the fringe area near Brahmandeo, *Hydrilla verticillata* are found in rainy season with *Najas*. Lots of *Gracillaria conforvoides*, a rhodophycaean algae are found on the rocky substratum of this zone.

Outer Channel Zone

It starts from Magarmukh area and end in lake mouth near village Arakhakuda. The water depth varied from a minimum of 1.5 to a maximum of 2.7 m. The salinity in this zone range from 4.5 to 31.7 ppt. In true sense there is no aquatic vegetation in this area. *P. pectinatus* is found in some blind pockets of the channel, where tidal effect is less pronounced. *H. oveta,* is the dominant plant, found on the sandy bottoms.

Recent Changes

The northern zone exhibits a varieties of hydrophytes. Pattanaik (1973) reported a few freshwater hydrophytes from this zone during his survey in the year 1962-64. He also reported that the plants like *A. pinnate, Pistia stratiotes, E. crassipes, H. verticillata, V. spiralis* used to enter the lagoon with flood water and disappeared after February and March, when water salinity increases. Besides these last two plants were much less. He also informed that during their two years survey period, he had not seen any flowering plant. But recently it is observed the flowering of *V. spiralis* & *E. crassipes* in the month of February, 1992. That means the plants, which were casual visitors to the lagoon earlier, permanently established in the lagoon during the period of past 30 years.

Biomas Macrophytes

Pattanaik (1973) reported biomas in terms of density in wet weight form from the whole lagoon. In the recent survey work, biomas of some hydrophytes has been

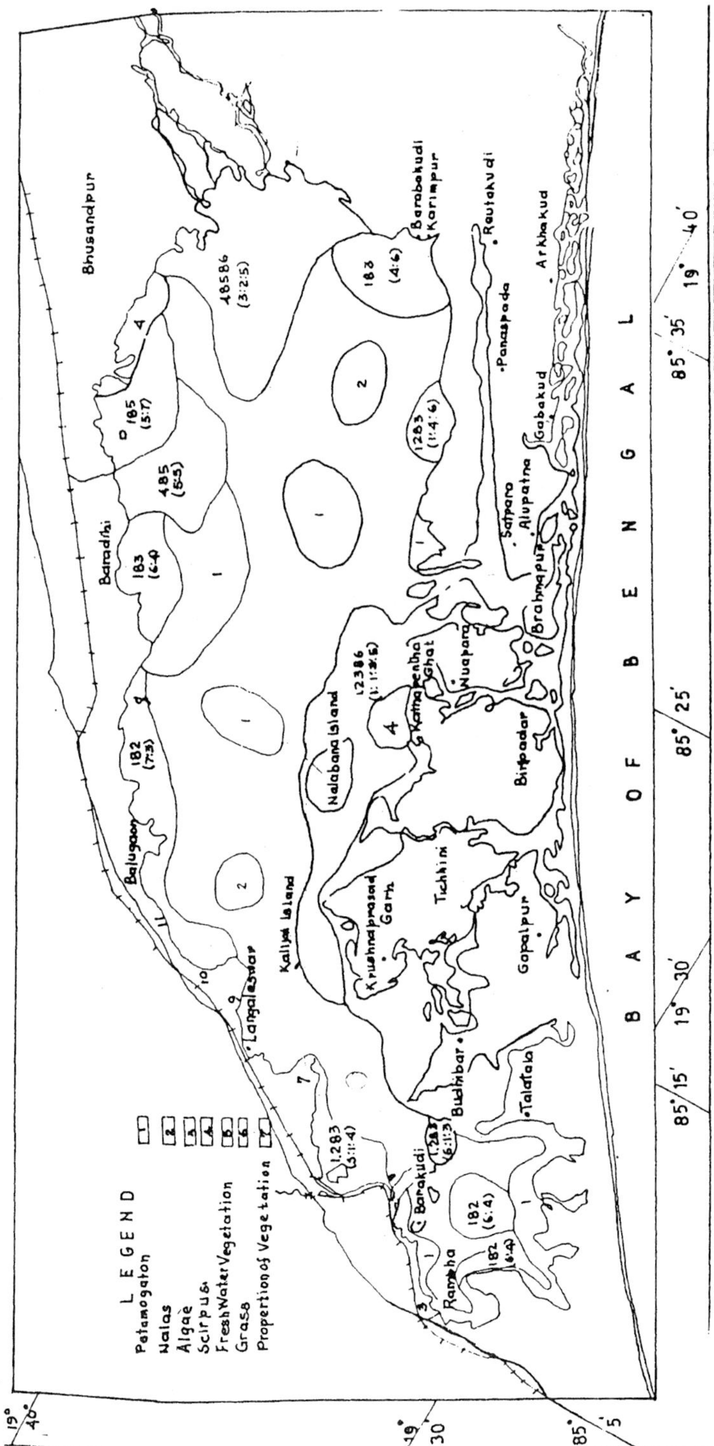

Flora of Chilka Lagoon

studied which was not reported earlier. The maximum wet wt. biomas was contributed by *E.crassipes* being 18,640 gms/m^2 (wet wt.) among the floating hydrophytes which include 8880 gms. of roots. In the softer sedimental bottom, where the water depth is low it has profuse root system. In the deeper region the root system is less. *A.pinnata* has been recorded as the lowest biomas among the floating weed being 80 gms./m^2 in wet weight. Among the submerged weed *P. pectinatus* recorded 4500 gms/m^2 (wet wt.) in the northern zone where the soft sediment in the bottom is more. But in the southern zone, only 3600 gms/m^2 (wet wt.) have been recorded at Panchbatia area near Parikud, where the sediment is less compared to the northern zone. Patnaik (1973) recorded the maximum biomas of *P. pectinatus* being 3780 gms/m^2 (wet wt.) *V. Spiralis* with recorded biomas of 7500 gms/m^2 (wet wt.) and *H.Verticillata* of 3740 gms/m^2 (wet wt.), was not recorded by Patnaik during his survey work. The emergent weed like *S. littoralis* with recorded biomas of 14,640 gms/m^2 (wet wt.) in the northern zone, includes 5560 gms. of root. Next to *S. littoralis, P. karka* recorded the high biomas of 5068 gms/m^2 (wet wt.) including root among the emergent weeds. The biomas of *Nymphea alba*, a floating macrophyte have 3140 gms./m^2 and 1340 gms/m^2 (wet wt.) of shoots and roots respectively. The grasses in Nalban (*P. repens* and *P. welence*) recorded biomas of 17520 gms/m^2 (wet wt.).

In 1973, small patch of about 20 sq. km. covered by *P. pectinatus* and *S. articulatus* community has been noticed in the north western zone. Dispersal of these communities were observed in the year 1977, where the colonisation have been established in two more areas, in addition to the initial one, totally occupying an area of about 60 sq.km. The growth of weeds in between the three axis of 1977 appears to have been completed in early eightees and in 1982, approximately 100 sq. km. of weed zone have been formed. During the period 1984-85 the weed community is seen closing in on the rest of water area and the weed area in these two years have been estimated to be about 180 and 200 sq. km. respectively.

In general the coverage of weeds over the years has been roughly calculated as 14.6 sq.km. per year and in 1984, 40 per cent of the lagoon water spread area were found to be covered by weeds.

The enormous and rapid growth of these weed communities amount to drastic shrinkage of water area of the lagoon leading to rapid eutrofication and reduced navigability in the long run if not checked immediately.

5

Commercial Activities in the Lagoon

Fishing is the single largest commercial activities of the lagoon.

The Chilka was owned by nearly one thousand owners, the principal ones being government, Raja of Parikud, Raja of Khallikote, Suna Bibi, Choudhury family of Bhingarpur, Mirza Taher Beg, Durga Charan Jagdev, Sridhar Khuntia, Jaganath Gochikar, Paikpara state. The Fisheries of central part of the lagoon (Bada Chilka) that remained full of water all year round, remained the exclusive right of the Government in pre-independence, while the waters of creeks and channels (Chhota Chilka) which becomes dry in summer, were owned by private estate holders, the biggest of them were Raja of Parikud and Khallikote. Fisheries in the areas belonging to their land holdings were leased out to private fish merchants, who used to sub-lease them to the fishermen (Bahanias) at a higher rate. The right of land holders over fisheries was later transferred to the Govt. under Orissa Estate Abolition Act.

The fishermen in villages adjoining the Bada Chilka had the right to fish, free of cost, until 1897-98, when a rent per house per year was introduced. This fee (Chulimunda tax) was raised to Rs. 1/- in 1923 and is still effective today for fishing in open waters by net.

The following fisheries have been identified and leased out in the lagoon; prior to Estate Abolition Act (Table 5.1).

Table 5.1

Sl. No.	*Type of fisheries*	*Total nos.*	*Owner*	*Lease value fixed (Rs.)*
1.	Traditional fisheries	18	Govt.of Orissa	5,400.00 (1 and 2)
2.	Prawn fisheries	6	Govt. of Orissa	
3.	Traditional fisheries	40	Raja of Parikud	37,902.00
4.	-do-	19	Mirza Taher Beg	10,270.00
5.	-do-	2	Mirza Taher Beg	600.00
6.	-do-	1	Gangadhar Choudhury, Bhingarpur	140.00
7.	-do-	1	Md.Basuruddin, Malud	180.00
8.	-do-	1	Jagirdar-Jagarjee, Malud	160.00
9.	-do-	1	Suna Bibi, Cuttack	750.00
10.	-do-	2	Durga Ch.Jagadev Roy, Puri	127.00
11.	-do-	2	Gopinath Gochikar, Puri	134.00

(contd.)

Sl. No.	*Type of fisheries*	*Total nos.*	*Owner*	*Lease value fixed (Rs.)*
12.	Traditional fisheries	1	Amir Alli	15.50
13.	-do-	1	Amir Bibi, Ganjam	5.75
14.	-do-	1	Gopinath Das, Bhinjarpur	90.00
15.	-do-	1	Sridhar Khuntia, Puri	5.25
16.	-do-	1	Padma Ch.Ray, Puri	13.75
17.	-do-	1	Raj Kishore Mohanty, Bentapur	27.50
18.	-do-	1	Braja Sundar Das, Bhingarpur	7.25
19.	-do-	1	Radhanath Das, Bhingarpur	3.25
20.	-do-	2	Lallababu, Estate of Paikpada	645.00
21.	-do-	1	Narsingh Mohapatra, Puri	18.00
22.	-do-	1	Padma Ch. Jena, Puri	250.00
23.	-do-	1	Kamal Bewa, Puri	50.00
24.	-do-	1	Mohan Mohanty	50.00
25.	-do-	2	Gambhari villagers	2,100.00
26.	-do-	1	Brahmachari Pattasing	800.00
27.	-do-	1	Bantamatha, Puri	1,000.00
28.	-do-	1	Priyanath, Mohanty	3,000.00
29.	-do-	1	Lingaraj Sahu, Bhusandpur	550.00
30.	-do-	1	Dasarathi Patnaik, Olasingh	100.00

A survey has been conducted in 1949 to identify fishing grounds of Chilka on the basis of general slope of the land, depth of water in the lagoon in different seasons and migration of fish in the area. As per survey, the fishing grounds have been classified into three categories.

Jano Fishing Grounds

A, one hundred ten (110) of such fishing grounds have been identified, though their numbers may fluctuate from year to year. These grounds are mostly located around the various islands of the lagoon.

During the period, a total of 110 janos were identified as exclusively leased out fisheries, contributing 13 to 22 per cent of total annual catches consisting of species like *M. cephalus*, *L. macrolepis*, *L. calcarifer*, *G. setifer*, *M. gulio*, *E. suratensis*, *Sparus* spp. and *P. indicus* and *P. monodon*.

During 1954-57, period, the Govt. of Orissa could collect an annual average lease value of Rs.70,695 by leasing 110 Jano fishing grounds.

Table 5.2

Sl. No.	*Name of fishery source*	*Location*	*Area (Ac.)*	*Lease value in 1991 (Rs.)*
		Puri Tahasil		
1.	Bahar Adanga Jano	Tua village	145.00	3260
2.	Panchabatia Chandabatia Jano	Sedua river	115.00	3260
3.	Dandinida Jano	Ratamati	35.00	3260
4.	Baharkanheipur Nala Jano	Baghamunda	4.92	3260
5.	Barapania Jano	Panchudora	680.00	3260
6.	Panchudora Jano	Barapania	275.00	3260
7.	Nuakhara Majhi Adanga Jano	Manikpatna	475.00	3260
8.	Bhobania Jano	Panaspada	649.66	3260
9.	Kanteighoda Sipla Jano	Gopinathpur	727.49	3260
10.	Garada Jano	Panaspada	380.02	3260
11.	Talagadagadia Jano	Badabenakudi	70.00	3260
		Krushnaprasad Tahasil		
12.	Marei Jano	Berhampur	795.72	3260
13.	Khitisahi Jano,	Khirisahi	235.00	3260
14.	Hatikhal Jano	Berhampur	200.00	3260
15.	Ginapatti Jano	Biripadar	98.84	3260
16.	Kankadagada Jano	Alandapatna	1003.34	3260
17.	Kankadagada Jano	Jamuna	927.58	3260
18.	Jayamala Jano	Chitrakote	677.02	3260
19.	Kananuturi Jano	Nuapada	84.42	3260
20.	Sankha Jano	Kurupal	546.28	3260
21.	Kampaua Jano	Biripadar	225.54	3260
22.	Tentadhua Jano	Nuapada	152.04	3260
23.	Diankudi Baliachatra Jano	Bajrakote	300.00	3260
24.	Amurtanidhi Jano	Alundapatna	303.52	3260

Prawn Fishing Grounds

Next in importance to jano fishing grounds and are located along the shore of the lagoon and in shallow areas around the islands. Important among them are

Nalabana, Barpania, Sherkuda, Ratamati, Khandakhai. The prawns have a preference for shallow water areas with muddy bottom with submerged vegetation. Due to creeping habit of prawns at the bottom in the shore regions, especially at night, they are mostly caught by traps set in those areas. The season of operation is from March to July.

A total of 67 leased out prawn fisheries accounted between 19 to 32% of the total annual prawn production of the lagoon. Thatta Kantha, a kind of pound trap and dhaudis and bozas were in operation to catch penaeid prawns from the lagoon.

Table 5.3

Sl. No.	*Name of fishery source*	*Location*	*Area (Ac.)*	*Lease value in 1991 (Rs.)*
1.	Kodiabatia	Badabenakudi	93.10	9638
2.	Dahikhiamuhan	Panaspada	21.35	9638
3.	Baharchilika	Barakudi	50.00	9638
4.	Kholandi	Badabenakudi	3.07	9638
5.	Jagadala Knetandi	Jagadol	21.00	9638
6.	Jagadal Golara	Golara	20.00	9638
7.	Haridas	Haridas	16.50	9638
8.	Serakhanda	Godikanda	85.00	9638
9.	Gadamarada	Tuanasi	20.00	9638
10.	Nuakhara Mujhiadanga	Manikpatna	475.00	9638
11.	Panchudora	Nuakhara Majhiadanga.	275.00	9638
12.	Seduanadi Niyatchara	Barapania	1762.00	9638
13.	Ratamati	Satapada Bankija	683.00	9638
14.	Basantapur	Khetandi	32.00	9638
15.	Dharanikudi-I	Dharanikudi	33.00	9638
16.	Dharanikudi-II	Dharanikudi	21.00	9638
17.	Chalisikhanda	Unsurveyed	1.20	9638
18.	Sukurapada	Berhampur	0.50	9638
19.	Kumarpur Bahar Chilika	Unsurveyed	0.50	9638
20.	Gaibata	Nuapada	0.50	9638
21.	Ostapalli	Mainsa	0.50	9638
22.	Ghimuda	Krushnaprasad	0.80	9638
23.	Jayamall	Krushnaprasad	0.50	9638
24.	Chdheyanasi	Krushnaprasad	0.50	9638
25.	Balinala	Biripadar	0.50	9638
26.	Budhiparpentha	Utarabahi	0.41	9638
27.	Kathapentha Bahar Chilka	Unsurveyed	0.50	9638
28.	Ganeswarnasi	Gurubai	0.50	9638
29.	Sankha	Nuapada	0.57	9638
30.	Chiradanasi	Ipping	0.50	9638

(contd.)

Sl. No.	Name of fishery source	Location	Area (Ac.)	Lease value in 1991 (Rs.)
31.	Panichua	Khirisahi	0.50	9638
32.	Mithakua	Mithakua	0.50	9638
33.	Aganasia	Unsurveyed	0.50	9638
34.	Jharakata	Jhanikuda	0.50	9638
35.	Abhayamuhan	Biripadar	0.50	9638
36.	Ginapalli	Biripadar	0.50	9638
37.	Panhari	Biripadar	0.50	9638
38.	Barunakhati	Barunakuda	310.72	9638
39.	Bodiakudi	Khatiakudi	0.50	9638
40.	Kothanikasi	Kandaragaon	0.50	9638
41.	Matiakudi	Gurubai	1.00	9638
42.	Khatisahi	Unsurveyed	0.50	9638
43.	Panibata	Gurubai	1.00	9638
44.	Garasara Pahanda	Tichhini	0.50	9638
45.	Khatiakudi Bahar Chilka	Bahar Chilka	0.50	9638
46.	Tampadakona	Rasakudi	0.50	9638
47.	Budhibar Pentha Dakhinbahi.	Unsurveyed	0.50	9638

Net Fishing Grounds

In the deeper areas of the lagoon, mostly gill nets are operated through out the year to catch fish.

Fishing off Satpada and in the outer-channel is done throughout the year on the location of fish shoals and favourable tide condition. Drag nets (Khadijal) and gill nets are operated in this area.

Fishing by seine-net is also done during the rainy season between Kaluparaghat and Sorana.

Between 1957-65, 39 net fishing grounds (Bahani) were leased out fisheries in the lagoon. With the exception of leased out Bahani grounds, any one is entitled to fish in unleased waters with payment of a nominal fee. Keuto fishermen used to do net fishing throughout the year with varying degree of intensity.

Net fishing contribute about 50 to 60 per cent of annual catch of the lagoon, dominating non-commercial fish catch (49.8 per cent) and fish like *P. coiber, M. cephalus, E. tetradactylum, N. nasus, M. gulio, H. illisha, P. indicus, L. calcarifer, O. militaries, L. macrolepis* and *G. setifer* among commercial fish.

The effective nets are patua jal, khadi jal and menjia jal.

Between 13 to 22 per cent of lagoon's total catches were obtained from janos, 110 of them covering 52.639 sq. miles were exclusively leased out fisheries in the lagoon during the period (1954-57).

Some smaller "jano" and "uthapani" are leased out directly by the Tahasildars, the local representative of the Revenue Department.

Table 5.4

Sl. No.	Name of fishery source	Location	Area (Ac.)	Lease value in 1991 (Rs.)
1.	Gangadharpur Nali	Padanpur	56.18	3260
2.	Raipur Nali	Padanpur	22.96	3260
3.	Chandramara Nadi-I	Chandramara	781.33	3260
4.	Chandramara Nadi-II	Chandramara	599.77	3260
5.	Gangadharpur Nadi	Manikpatna	310.34	3260
6.	Gorapur Nadi	Manikpatna	213.08	3260
7.	Parbatipur Nadi	Manikpatna	75.40	3260
8.	Sanapatna Nadi	Manikpatna	581.30	3260
9.	Gabakunda Nadi	Manikpatna	766.44	3260
10.	Arkhakuda-I Nadi	Manikpatna	774.70	3260
11.	Arkhakuda-II Nadi	Manikpata	887.93	3260
12.	Arkhakuda-III Nadi	Manikpatna	654.82	3260
13.	Marei	Berhampur	1059.15	3260
14	Manisa Gramtali-I	Gothakuda	1000.00	3260
15.	Manisa Gramtali-II	Gothakuda	1000.00	3260
16.	Manisa Gramtali-III	Gothakuda	550.00	3260
17.	Khatisahi Gramtali-I	Barunakhati	919.00	3260
18.	Khatisahi Gramtali-II	Barunakhati	1000.00	3260
19.	Jalimuhan Chikhal Khandi-I	Budhibor	1000.00	3260
20.	Jalimuhan Chikhal Khandi-II	Budhibor	825.00	3260
21.	Bahar Chilika	Ramalenka	Unsurveyed	3260
22.	Bahar Chilika	Nuapada	Unsurveyed	3260
23.	Bahar Chilka	Biripadar	Unsurveyed	3260
24.	Bahar Chilika	Jhilimithakua	Unsurveyed	3260
25.	Mathapur	Tentulipada	Unsurveyed	3260
26.	Pathura Tolasahi	Chandraput	Unsurveyed	3260
27.	Bankiapalli	Patanasi	Unsurveyed	3260
28.	Baliapatpur	Baliapatpur	Unsurveyed	3260
29.	Bhusandapur	Bhusandapur	Unsurveyed	3260
30.	Pathara Uppersahi	Pathara	Unsurveyed	3260
31.	Baradi	Baradi	Unsurveyed	3260
32.	Jaripada	Jaripada	Unsurveyed	3260
33.	Kumandalo	Kumandala	Unsurveyed	3260
34.	Nairi	Nairi	Unsurveyed	3260
35.	Baulabandha	Baulabandha	Unsurveyed	3260
36.	Dhuanala	Dhuanala	Unsurveyed	3260
37.	Kaudikhani	Kaudikhani	Unsurveyed	3260
38	Gourangpatna	Gaurangpatna	Unsurveyed	3260

Uthapani (fishing in shallow areas during first rainfall in June).

Table 5.5: Puri Tahasil

Sl. No.	*Name of Uthapani fishery source*	*Location*	*Area (Ac.)*	*Lease value in 1991 (Rs.)*
1.	Kalandikuda	Badabenakudi	9.18	3260
2.	Kholanala	Badabenakudi	3.07	3260
3.	Kaliaghai Kantia Uthapani	Badabenakudi	63.12	3260
4.	Badanala	Baghamunda	77.55	3260

Dian (fishing in small shallow areas upland from some Jano-areas.)

Table 5.6: Puri Tahasil

Sl. No.	*Name of Dian fishery source*	*Location*	*Area (Ac.)*	*Lease value in 1991 (Rs.)*
1.	Jagadal Golara	Jagadal	42.00	3260
2.	Jagadal Golara Bahar Chilika	Jagadal	90.00	3260

Fishing Gears & Methods

There are 54 synonyms used in local fishing gears of the lagoon to denote only 14 well defined gears depending on the following fishing methods adopted in Chilka.

(a) Jano fishing
(b) Net fishing
(c) Trap fishing
(d) Harpooning

Jano Fishing

Jano—a community fishing, has a long tradition in Chilka, enclosing fish in shallow water areas by bamboo screens. To-day, due to very shallow water close to land, the major part of the enclosement is made of low mud bundhs. Bamboo screens are used only in some openings of the mud wall.

Lots of low-lying land become covered up with water during post-monsoon period of 3-6 months. These areas become highly productive, due to its shallowness with the influence of sunlight and high temperature. With the slight rise of salinity in September-October, many fish species migrate in those shallow water areas, which are been watched by the fishermen. When the fish come to the shallow areas, they enclose it, very quickly with the help of 70-100 fishermen.

In little deeper water only split bamboo screen are used to close the Jano in a few minutes with the help of 30-40 nahas. The Janos are operated from October to February by a group of fishermen, namely Kandara, the second largest fishing community of the lagoon. The first Jano to be closed are situated in the southwest and then the other janos follow in a north-easterly direction.

Depending on the size of the jano, it has one or more catching areas, called gohoras, made up of two parallel bamboo or net screens about 3 metres apart. On the inner screen there is a small opening, through which the fish swim against the current and finds its way out inside the gohoras and unable to return back into the jano. The gohora is then emptied with the help of sticknets (khadijal) or bamboo mats, usually once a day, but 2 or 3 times if there are lot of fish. The small sized fish from gohora is thrown back into the jano to grow more. Upto 12 persons depending on the size stay at the site of the jano, guard catch and repair when needed.

During main season, all the catching takes place in the gohora and some times in a few prawn traps inside the jano. At the end of the season, the fishing activities also move inside the jano and the remaining fish are caught by net fishing.

As the season nears its end, most janos have again become dryland.

The biggest jano, Nalaban has been declared as bird sanctuary since 1988.

Different species dominate the catch in different janos. In some, 80 per cent of the catch could be of one species.

The dominant species caught in the janos are *Mugil cephalus, Liza macrolepis, Lates calcarifer, Gerres setifer, Mystus gulio.*

Equipments for jano fishing are made out of plants, wood fibres, saps, leaves with some synthetic net materials introduced recently. To close a jano screen, pointed sticks made from bamboo (*Bambusa arundinaceae*) or casurina of 3 m. length and boats are needed. Screens (thattas) vary in length 5-10 m. and height 1.2-2 mt. tying the bamboo splints together with creeper plants.

Materials used for jano operation are:

(i) Thatta—Bamboo mat.
— mesh 8 mm, 2m high, 10 mt. long.
— mesh 5 mm, 1.5m. high, 5 m. long.
— mesh 5 mm, 2.2-2.5 m. high, 10m. long.

(ii) Bada jal—Leader line, bamboo.
— 1 m. high, 5 m. long.
— 1.2m. high 3.3 m. long.

(iii) Khadi jal—Sticknet, nylon and bamboo stick
— 0.6-0.8 m. apart.

Dhaudi — Big bamboo prawn trap.

Disco Dhaudi — Big HDPE netting cage prawn trap
— $1 m^3$.

Wooden sticks—Sticks for the jano enclosures.
— 3 m. high, bamboo or casurina.

Katha — Punting stick-bamboo

Sail — Bamboo mat.

Between 13 to 22 per cent of lagoon's annual catches are accounted by Janos and the production from a normal Jano was estimated to be 124 Kg./ha. per season.

The total cost of materials for an average jano is hard to get. For the Khani Muhan jano (450 ha.), the cost of materials was Rs. 20,000. This was for the raw-material. The fishermen manufactured the screens themselves. For the other jano, Solakudi (165 ha.) the cost was about Rs.9,000. Based on these two figures the estimated cost of materials should be between 45 and 55 Rs/ha. (Nov.88).

Other costs include licences, commissions and lease-fees. A licence is needed to operate a boat in Chilka. The licence costs between Rs. 30 and 50 and the yearly fee is Rs. 10-20 depending on the size of the boat.

Net Fishing

Various fishing nets, which are in operation in Chilka are as under:

(1) Drag Nets
- (a) With bag, wings and a cod end.
  - (i) Bekti jal
  - (ii) Bhida jal
  - (iii) Patua jal
- (b) Without bag
  - (i) Boroga jal
  - (ii) Khadi jal
  - (iii) Mani jal
  - (iv) Patua jal

(2) Gill Nets
- (a) With foot rope
  - (i) Dosti jal
  - (ii) Menjia jal
  - (iii) Ora jal
  - (iv) Sahal jal
  - (v) Hilsa jal
- (b) Without foot rope
  - (i) Noli jal (drift net)
  - (ii) Khoinga jal
  - (iii) Chaugunda jal
  - (iv) Dosta jal
  - (v) Dosti jal
  - (vi) Dalua tir jal
  - (vii) Orandhi jal
  - (viii) Tir jal
  - (ix) Kekenda jal

(3) Cast Nets
- (a) With strings
  - (i) Khepa jal
- (b) Without strings
  - (i) Khepa jal

The net fishing contributes nearly 50 to 66 per cent of annual fish catch from the lagoon. Many of the nets are more selective in respect of particular fish species after which they are named. The net-fishing intensity varies during the year and is low during October-December coinciding with jano season.

Non-commercial fish dominantly occur in net catch forming almost 50 per cent and the dominant species in net catches are *Pseudoscinna coiber*, *Mugil cephalus*, *Eulethrenema tetradactylum*, *Nematolosa nasus*, *Mystus gulio*, *Hilsa ilisha*, *Penaeus indicus*, *Lates calcarifer*, *Osteogeneosis militaries*, *Liza macrolepis*, and *Gerres setifer*.

Major type of fishing net operating at the lagoon are—

(1) *Khepa jal*— (Cast nets) Height 3-6 m, circumference 12-27m. mesh size 12-25 mm) operated by Keuto, Kartia and Gokha fishermen of the lagoon. These nets are operated in shallow waters by whirling them round the head like a lasso, and are usually cast when the movement of fish shoals have been located by the operator. At times cast nets are operated in group fishing by twenty or more fishermen. On spotting a shoal, about 10 to 12 large nets are spread in an acute angle from the shore and the fish driven to this area by other fishermen in boats. Small fishes are usually captured by cast nets.

(2) *Patua jal*—A seine-net to catch small fish as well as bigger varieties like hilsa (*H. ilisha*), sahal (*E. tetradactylum*) are operated by the fishermen of the northern and western shores from Balugaon till Daya river from November to April.

(3) *Bekti jal*— Large gill nets, now made of synthetic twine, are used only in the "Gohiras" or deeper portion of the lake to catch bekti (*L. calcarifer*) sahal (*E. tetradactylum* and borog (*P. coiber*) etc.

(4) *Bhida jal*— Drag net with bag (length-7.5 m, Breadth 2.7-3m. mesh 19 mm) fitted with head rope and 10-15 floats to catch hilsa, thread fins, rays, pearl spot, sharks.

(5) *Noli jal*— Small meshed gillnet of 15-18 m. length and 0.9-1.05 m depths operated chiefly by the fishermen of the north-western areas of the lake for catching small mullets, medium sized fish and crabs, using small thin bamboo sticks (90-100 m long) as floats.

(6) *Khadi jal*— A drag net ranging from 10 to 15 m. in length with various depth and 6 to 19 mm. mesh are used all over the lagoon, but chiefly in the outer channel and neighbouring areas. The opening of the net is kept open by a number of bamboo sticks of different lengths set about 0.6m. apart. Different varieties of fish of assorted sizes are caught in the net depending on the mesh size.

(7) *Muni Jal*— The light surface drag net, specially meant for catching beloniform fishes are used by the Keuto fishermen of Chandraput and Pathara villages.

(8) *Hilsa jal*—Gill net used for catching hilsa near the mouths of river Daya and Bhargavi by the fishermen of villages situated on the Daya.

(9) *Gohira jal*— Used by some of the fishermen of the south-western area for catching fish in the deeper portion of the janos, measuring 5x4 m. and now made of synthetic twine.

(10) *Pellona jal*— Pushing nets used to catch small fish and prawns.

(11) *Paramanand jal*— A gill net of 27 m. long, 1.3m depth with 75 mm mesh are operated by fishermen of Chandraput for catching bhekti. 10 such nets are joined together to encircle a large area.

Lifting Gill-net in Chilka Lagoon

(12) *Menjia jal*—A gill net of 7 m. length and 3.5 m. breadth with 13-19 mm. mesh are in operation in shallow water area off Balugaon, Parikud to Bhusandapur during summer. While two ends of the net are held from two boats from both the sides, 7 to 8 fishermen scare the fish to it. After scaring the fish, the two ends of the net joined together and the net is gradually hauled in, Smaller mullets, hilsa, thread fins and flat fishes are caught in this net.

(13) *Kapa jal*— The net , a modified Khadijal of 14 m. length, 1 m. breadth and 6 mm. mesh and 22 spreader sticks are operated in shallow water areas between Krishnaprasad and Pathara to catch beaked fishes. The Kondras of Pathara, Langaleswar, Sabelia, Keshpur and Barkul own this net and use it in day. Six persons hold five nets in a semicircular fashion, one hundred metres off from the shore with the opening of the semi-circle towards the shore. Meanwhile two people draw a scare line from the shore towards the net splashing the water to scare the fish into the net. As the scare line comes very near the semi-circular net formation, the nets are closed up and the circle is completed quickly and the group is divided into three units to form the smaller circles, after which spreader sticks are pulled very quickly.

(14) *Modhi jal*— Twenty fishermen from Balugaon to Bhusandapur area operate this net of 4 m. long, 3.3 m. depth and 38 mm. mesh, similar to that of behkti jal, in winter and early spring, when there is no strong wind. Sorana fishermen generally use this net to catch bhekti, cat fish and scianids.

(15) *Olandi jal*— A gill net of 65 m. long, 0.6 m. depth and 6 mm. mesh is operated in shallow water of Noiri area to catch small mullets, cat fishes, flat fishes etc. The net can be operated by one man, who simply spreads the net in shallow water and scares the fish to it for sometimes, after which the net is dragged. But usually two people operate this net.

Sea-nets used by the Nolia fishermen of the Chilka area include Bada-jal, Irrigali and Kavalo jal.

In recent years, displaced fishermen from Bangladesh operate boat seine of 50 m. length and 17m. depth with 17-22 mm. mesh size in the open lagoon to catch all types of fishes. The net is operated with the help of 5 to 7 boats with 25 to 35 fishermen.

Trap Fishing

Three types of trap are in operation in the Chilka lagoon of which large prawn trap made of split bamboo locally known as Dhaudis are set in shallow waters by the Kandara fishermen to catch prawn. Smaller prawn traps (Bozas) are operated chiefly by Tiara fishermen of the lagoon. Besides "Kankada Khadia" a rectangular box type bamboo traps are used in large numbers to catch crabs.

Thattas or bamboo screens of various size in different areas are used as leader lines with the prawn traps. These bamboo screens are also arranged to serve as fish traps without any "Bozas" or "Dhaudis". Small thattas are locally known as "Naharas".

Bamboo Traps

(1) *Dhaudi*— A busket trap of box type with a conical top. The height is 1.2 m.

Mullets Caught in Gill-net — A Poor Catch

and length 1.15 m. with 0.4 m. bottom. There are two to three openings at one side having thin bamboo strip intersections traps valve through which prawn can enter inside but can not come out.

Five to ten such traps are arranged in a circular manner leaving a narrow opening from which the leader line made of strips of bamboo (thattas) emerges extending upto the shore. Prawns coming in contact with the leader line follow the line and enter into the trap. 100 such traps arranged in rows without any leader line have also been observed in some areas. Dhudies, a bigger one than Baja are operated in deeper waters, while Baja are operated in shallow waters.

(ii) *Kankada Khadia*— A box type crab trap, made up of bamboo strips of 6 mm. width separated with inter-space of 12 mm. and dimensions of 0.45 m x 0.3 m. X0.25m. is operated to catch crabs (*Scylla serrata*). In the side of 0.3 m. x0.25 m., a slit opening of 20x10mm. is provided covered by means of 'V' shaped bamboo splits projected inside the trap.

Baits of small fish or prawns is kept inside the trap to attract the crabs and the trap is kept at the bottom tied to a pole, which are being checked at an interval of 2-3 hours to collect the entrapped crabs.

(iii) *Disco net trap*— With the switching over to HDPE twine from natural fibre nettings and split bamboo traps, due to increase in their price, a box type net trap, known as "Disco-net trap" has replaced the bamboo traps and thattas and is

Fixing Baja (Trap) with Leader Line for Catching Prawn

popularised in the lagoon for its comparatively low cost and more catching efficiency.

The unit consists of two boxes (poda) of 1.6x1.6x1.6 m dimension made up of HDPE netting of 21-25 mm. mesh size with a central leader line (Bada) between the boxes running for 116 m. long of HDPE netting supported by sticks fixed on the ground. Two HDPE leader line (Binchini) consisting of 750 meshes, (Danda) of 21-25 mm. size run along the back of the boxes in both the sides. A total of 5.25 kg. of HDPE netting is required to make a unit and the cost per unit comes to Rs.500/-. The trap net is set in the evening mainly for prawn and lifted in the morning. 2-3 kgs. of prawn are caught by a unit per operation.

Hook & Lines

A long line of 70-100 m. attached with hooks at an interval of 3 to 3.5 m. is stressed across the water with one end tied to pole and the other with the boat. Rotten prawn and fish are used as baits. This fishing method is adopted near Satpada and Baruna and in areas between Nuapada and Parikud. Catfishes are found to be caught in hook and lines.

Harpooning

Spear fishing is practised in the lagoon during dark nights. A burning lamp is so hung from the fore-deck of the boat that it remains just above the water level. A man holding four pronged iron spear attached to bamboo pole measuring 3 m. stands at the end of the fore-deck, while another noiselessly punts the boat from rear end. The man at the fore-end spears the fish as it comes close to the boat, being attracted with naked light. Chiefly mullets (*M. cephalus*) and *L. macrolepis* are speared in large quantities.

Fishing Crafts

Dug-out (donga) and flat bottomed boat (Naha) are in operation in Chilka lagoon. The flat bottomed boat vary in three sizes namely big, medium and small and are made sal wood (*Shorea robusta*) plank of 25 to 38 mm thickness. Their detailed specifications are as shown in Table 5.7.

Table 5.7

	Size	*Length overall in metres*	*Maximum breadth in metres*	*Depth at mid-ship in metres*
(a)	Large	13.5-15.0	1.90-2.20	0.70-0.85
(b)	Medium	12.0-13.5	1.60-1.80	0.60-0.68
(c)	Small	5.5-10.20	1.20-1.30	0.44-0.50

Most of the equipments for the boat are made out of bamboo (*Bambusa-arudinaceae*) such as sail, rudder, handle, masts, punting sticks.

Large, rectangular sails, made from long but thin bamboo mattings and supported at short interval by stout bamboo pieces are attached to the flat-bottomed boats. They are rolled and kept away, when not in use.

Boats Operated in Chilka lagoon

Table 5.8: The Boat Type Used in Chilka Fisheries

Name (local)	*Length in metres*	*Breadth in metres*	*Depth in metres*	*Present Cost in Rupees*
Bautia	9.5-10.2	1.2-1-3	0.44-0.5	8,000-10,000
Donga (medium)	12.0-13.5	1.6-1.8	0.60-0.68	20,000-25,000
Naha	13.5-15.0	1.9-2.2	0.70-0.85	30,000-35,000

Bautia are used for off-shore fishing with boat-seine to catch small fishes like pearl spots etc., while medium sized donga and naha are used for gill-netting and net fishing (bahni) and large sized ones for transportation of fish, fishing materials for jano fishing. The speed of these boats depend on favourable winds. In favourable wind with a sail they can move in 40 kms. per hour, while in absence of wind , the boat has to cruise with punting sticks in a speed not more than 4 kms. per hour.

Medium and large sized donga and naha has recently been motorised with the help of outboard motor of 7.5-11.5 horse power operated by either diesel oil or kerosene. With the motorisation their speed increased to 15 to 20 kms. per hour irrespective of wind.

Nolia fishermen possessing crafts like catamarans and Podhua (sea going boats) use them only for sea-fishing.

6

Traditional Fishermen—Caste and Status

Chilka lake has 122 fishing villages with a total population of around 85,000 not including the surrounding area which has about 273 villages. About 30 per cent (27,200) of the fishing village population are active fishermen, although many others depend indirectly on the fisheries. The specific methods of fishing are a reflection of the particular caste characteristics of the community.

Three distinct communities can be identified as having crucial linkages to the lake and its resource management.

(a) the fishermen (traditional and non-traditional),

(b) the farmers who live around the lake, and

(c) those who depend on the forest resources in the lake catchment area for both their livelihood and to meet their fuel/timber requirements.

The traditional fishermen are Harijans the "Untouchables" and hence they occupy the lowest social position in the society. In addition, their poverty continues to reinforce this lower social position as, in the changing Indian village scenario, social status and economic power go hand in hand.

Fishing labour of the lagoon (fishermen) belongs to Keuto, Nairi, Kartia, Gokha, Mandra and Tiara castes among Oriyas and Nolia, living only on the sea coasts are Telegues.

Most of the fishermen, who are illiterate and poor, do not seem to have any ambition. Their poverty restricts their fish food to small and cheap varieties, while the high priced varieties are sold. Their scanty clothings and general low standard of living is primarily due to their poverty, which has not risen inspite of the increase of catches in recent years.

A survey in 1957 undertaken by the State Fisheries Department revealed that out of total population 31,550, living in and around the lagoon, 21375 (67.1 per cent) were Keutos, who are considered to belong to the higher stratas among fishermen. These fishermen catch fish only with nets and are not supposed to use traps, which are meant for lower fishermen castes. However, some of the fishermen of Satpada village were reported to use Dhaudis for catching prawns. Keutos are distributed mostly near the Satpada area and on the northern and western shores of the lagoon.

Nairi— A small group of fishermen, 1.9 per cent of total population, very similar to Keutos in social status were confined to a few fishing villages situated on the western shore of the lake. They operate nets for fishing and do not use traps. This group are found gradually changing over to other professions, such as farming, commission agents for collecting fish from fishing grounds to the assembly points.

Prawn Catch from the Lagoon with Bajas

Kartia— A minor fishermen caste contributing 2.7 per cent of the total population, living in five villages, namely Barakudi and Baghananji in the north-east, Panchupatia and Mangalajori in the north-west near Kaluparaghat and Kumandal near Gangadharpur. The kartia of two former villages and kumandal operate bamboo screen traps for fishing, while those of Mangalajori use traps and nets and those of Panchupatia only nets.

Gokha— About 0.2 per cent belonging to the inferior caste live in two villages in the north east. They operate drag nets and cast nets for fishing.

Kandra— The second largest fishing community of the lagoon comprising 14.3 per cent of the total fishermen population and are distributed mostly on the south-west and south-east shores. Jano fishing is mostly done by kandaras. Those who operate traps like Dhaudis and Thattas together with the Tiara share the monopoly of the catch (prawn & crabs).

Tiara— Constitute 7 per cent of the total fishing population. They are distributed in the north-eastern and north-western shores and fish with Bozas and Thattas.

Nolia—These Telegu fishermen, fishing, mainly in the sea and partly near the south and south-eastern portions of the lagoon constitute 6.8 per cent of the total. Nolia villages are situated between the sea and the Chilka in the south-eastern shore, who at times fish in the lake with drag nets.

There is significant status differentiation between the subcastes, primarily related to economic status. But between Khatia and Tiara, there are only marginal differentiations. While Khatias claim that they occupy the highest position, others challenge this. Kandras are the lowest of the subcastes.

The illiteracy of the fishermen is partly due to the elders, who believe that regular attendance of the children at school would deprive them of the assistance rendered by the children during fishing operations. They also apprehend that education may wean the children from the traditions of the communities. The fishermen take pride in fishing in the areas, where their fore-fathers fished for generations.

Social contacts of the fishermen are confined to religious festivals, marriage functions etc., when fishing is stopped. Rural amusements are arranged during these occasions. The Goddess Kali at Kalijai is the chief diety worshipped by the fishermen of the lagoon. A festival on Makar Sankaranti, which usually falls during January is a great day at Kalijai. Dol purnima, Ekadasis are other days observed by them.

A total of 2351 boats, 49985 nets and 81028 traps have been reported from 114 fishing villages in and around the lagoon during 1957 as detailed below—

Large boats	617
Small boats	1734
Khepa jal	1566
Patua jal	13417
Bhekti jal	12467
Bhida jal	9063
Noli jal	5541
Khadi jal	5431

Muni jal	428
Hilsa jal	1557
Other nets	252
Sea fishing nets	263
Dhaudi	38309
Boja	18665
Thatta	24054

Recent surveys of Chilka lagoon area reveals that out of 273 villages situated in the periphery and in the islands of the lagoon, 128 villages are dominated by fishermen castes, having 12363 fishermen families with a total population of 85,037 as detailed below—

Males	25,039
Females	21,693
Children	38,305
Total	85,037

22,095 males are reported to be actively engaged in fishing in the lagoon.

Besides a big group of refugees from Bangladesh has settled in Chilka near Balugaon. They are about 8400 in numbers and are making their living mostly out of fishing, 1500 of them are active fishermen.

Recently other groups of people have moved to Chilka. Even people of Brahmin caste have come to use water areas for the profitable prawn farming.

A sample survey of fifty one villages on Chilka coast, by ORSAC during January to March 1988 discloses the following facts:

1.	Total population in 51 villages		25,275
2.	Total population of Sch. castes		12,574
3.	Total population of other castes		12,701
4.	Population of male fishermen		6,493
5.	Population of fishermen		5,864
6.	Active fishermen engaged in fishing		
	Adult male	3560	
	Adult female	168	
	Children	529 =	4,257
7.	Population of immigrants since independence upto 1987		1,234
8.	Population of immigrants since last five year		932
9.	Total land possessed by fishermen		259.13 ha.
10.	Total land possessed by other castes		1156.27 ha.
11.	Literacy rate of fishermen		22.09%
12.	Literacy rate of other castes		29.75%
13.	Average per capita income of all these villages from fishing		Rs.500.00
14.	Average per capita income of all these villages from agriculture		Rs.155.00
15.	Average per capita income of all these villages from other sources		Rs.352.00

16.	Total land of fishermen for cultivation	225.2 ha.
17.	Total land of others for cultivation	1294.6 ha.
18.	People engaged in cultivation	1574 families (9900 population)
19.	Total number of boats	1064

The use of different parts of the lagoon depending on the technique used has led to the pattern of settlements for the different castes adopting different fishing methods around the lake.

This stratification of the population into groups using different technique for utilising different resources, has made possible for more people (60 thousand people directly or indirectly depend on the lagoon for their daily lively hood) to use Chilka as a source than if everyone had taken to the same technique. It is like the fishing population had become part of a quite stable ecological system. Because of the population increase some people have been forced to move or change occupation not to disturb the stable state of the Chilka ecosystem. This was the way how it was in previous days, but to-day much of these intricate system has changed and it is creating problem in many different ways, the major among them are—

(a) The population increase, both from nativity and influx of refugees, leading to more active fishermen in the lagoon.

(b) The catching or producing as much prawns are possible due to price increase since prawn export began.

(c) The physical change of Chilka caused by siltation, weed infestation etc.

In present time, the people involved in fishing profession in Chilka can be grouped into following three categories—

1. Fishermen (caste) following fishing profession with negligible land holding for agriculture.
2. People of other castes, who are native of Chilka periphery or settled since ten years engaged in fishing, owing to low income from agriculture and they prefer bonded labour to fishing profession.
3. A large number of immigrant fishermen who come to Chilka for fishing only to accumulate much wealth in least labour.

The fishing craft and gear do not belong to the fishermen in some cases for which they have to give a good portion of their income towards rent of the equipments. But a significant portion of fishermen population, who are either native or immigrants since ten years are not facing such type of difficulty. They have purchased boats and other fishing equipments of their own.

As regards the community welfare of the people living in and around Chilka, the following facilities have been rendered to them.

1.	Colleges	2
2.	Higher Secondary high schools	48
3.	Middle schools	163
4.	Primary schools	630
5.	Hospitals	3
6.	Primary health centres	13
7.	Dispensaries	19
8.	Maternity centres	7
9.	Family planning centres	5

Besides fishing, the following fishermen and fisherwomen are reported to be engaged in other ancillary activities as detailed below—

1. Marketing of fish 3121
2. Net weaving 3013
3. Basket making 875
4. Other works relating to fishing 1413

Recent survey indicates the operation of 207 motorised boat and 5171 non-motorised boats in Chilka. A total of 19190 fishing nets of different kinds are used for fishing in the lagoon.

The fishermen still live separately from the caste Hindus. Some ten to fifteen years ago, the fishermen were not even permitted to enter the caste-Hindu villages and untouchability prevailed. Things have since changed significantly: the fishermen take baths in the village open pond where caste-Hindus also take baths and celebrate certain village temple festivals together. However, in a few places, some discriminations still prevail. For example, drinking water wells are separate; separate vessels are kept for them in the hotels, they have to carry the bride for the marriage, play traditional music, etc. Social awareness, education and economic development are crucial aspects which can enhance their social status in the society.

7

Fish and Crustacean Ecology—Fisheries of the Lagoon

Fish caught from Chilka lagoon serve as an important resource within the regional economy besides providing a source of high quality protein for local consumption. Chilka fish are exported to markets (largely in Calcutta) providing a significant source of revenue for numerous participants in the fishery. Due to current export potential and high prices obtained for landed prawns, they continue to maintain an important position in the commercial fishery within the lagoon.

Fish and crustaceans landings from Chilka lagoon from 1929-30 to 1993-94 alongwith their trend values (Five years moving average) is given in Table 7.1.

Table 7.1

Year	*Landings of fish and prawn*	*Trend values (5 years moving average)*	*Landings of crabs*	*Trend values (5 years moving average)*
1929-30	1730	—		
1930-31	1440	—		
1931-32	1150	1452		
1932-33	1510	1390		
1933-34	1430	1474		
1934-35	1420	1570		
1935-36	1860	1616		
1936-37	1630	1730		
1937-38	1740	1808		
1938-39	2040	1724		
1939-40	1770	1617		
1940-41	1440	1629		
1941-42	1095	1755		
1942-43	1800	1725		
1943-44	2670	1929		
1944-45	1620	2434		
1945-46	2460	2876		
1946-47	3620	3078		
1947-48	4010	3873		
1948-49	3680	4349		

(contd....)

Year	*Landings of fish and prawn*	*Trend values (5 years moving average)*	*Landings of crabs*	*Trend values (5 years moving average)*
1949-50	5415	4482.4		
1950-51	5020	4741		
1951-52	4287	4900.6		
1952-53	5303	4552		
1953-54	4478	4505.6		
1954-55	3672	4577.6		
1955-56	4788	4658.4		
1956-57	4647	4706		
1957-58	5707	4882		
1958-59	4716	4574		
1959-60	4552	4350		
1960-61	3248	4103		
1961-62	3527	4060		
1962-63	4472	3680.6		
1963-64	4501	3632.8		
1964-65	2655	3625.6		
1965-66	3009	3251.4		
1966-67	3491	3058.8		
1967-68	2601	4304.4		
1968-69	3538	5116.6		
1969-70	—	—	—	—
1970-71	—	—	52.9	—
1971-72	—	—	67.9	—
1972-73	8883	5799	64.5	64.4
1973-74	7070	6373.4	66.7	67.0
1974-75	6903	6698.6	70.0	66.0
1975-76	5473	6179.6	66.0	64.5
1976-77	5164	6042.8	63.0	64.8
1977-78	6288	5696	57.0	62.6
1978-79	6386	5879	68.0	73.9
1979-80	5169	6339.4	59.2	77.3
1980-81	6287	6421	122.5	83.5
1981-82	7466	6593.8	79.9	98.2
1982-83	6696	6759.8	87.9	104.3
1983-84	7250	7220.2	141.4	95.6
1984-85	5999	7501.6	89.9	90.4
1985-86	8589	7857.4	78.7	80.6
1986-87	8873	7626.6	54.0	61.0
1987-88	8475	7760.8	39.1	50.2
1988-89	6096	6897.6	43.5	39.1

(contd....)

Year	*Landings of fish and prawn*	*Trend values (5 years moving average)*	*Landings of crabs*	*Trend values (5 years moving average)*
1989-90	6670	6011.8	35.6	34.3
1990-91	4273	5148.6	23.6	29.5
1991-92	4545	4595.2	29.6	23.2
1992-93	4159	—	15.4	—
1993-94	3329	—	11.9	—

Landings of fin fish and prawn together and of mud crab (*Scylla serrata*) have been shown separately. The controversial landing data of fish and prawn of three years (1969-70 to 1971-72) have not been taken into account. Informations on crab landings earlier to 1970-71 are not available.

The trend values of fish and prawn from 1931-32 shows gradual increase from 1452 tonnes and thereafter dropped by 40 to 300 tonnes each year till 1963-64 when the landings stabilised in 3132 tonnes in 1963-64 and 3625 tonnes in 1964-65. The catch further dropped by 200 to 374 tonnes each year and finally lowest catch of the decades (3058 tonnes) were recorded in 1966-67.

In 1967-68, the catch suddenly rose to 4304 tonnes, a difference of almost 1246 tonnes from that of previous year and in 1968-69, the catch again increased to 5116 tonnes, showing a rise of 812 tonnes from that of 1967-68.

From 1972-73 to 1982-83, the annual landings varied from 5696 tonnes in 1977-78 to 6759 tonnes in 1982-83. From 1983-84, the landings showed an annual increase of over 300 tonnes, with the highest catch of 7857 tonnes in 1986-87 recorded in past six decades under report. The catch dropped in 1986-87 by 231 tonnes and again increased by 134 tonnes in 1987-88. In 1988-89, the catch again dropped by 863 tonnes giving the trend value of 6897 tonnes in 1988-89.

The gradual decreasing trend in catch continued in succeeding years when the lowest trend value of 4595 tonnes were estimated in 1991-92. The annual landings further dropped to 4159 tonnes in 1992-93 and 3329 tonnes in 1993-94.

The landing of mud crab (*Scylla serrata*) in 1970-71 was 52.9 tonnes. The catch increased thereafter annually with marginal variations, till it reached 122.5 tonnes in 1980-81. In 1981-82, the landings dropped to 79.9 tonnes but subsequently rose to 87.9 tonnes in 1982-83 and highest of 141.4 tonnes in 1983-84. Thereafter progressive decrease in crab landings were noticed in subsequent years, till it reach the lowest value of 11.9 tonnes in 1993-94.

From the trend values (5 years moving average), progressive increase from 64.4 tonnes in 1972-73, have been noticed with marginal variations when the highest value of 104.3 tonnes were recorded during 1982-83. From 1983-84, onwards, the landings of crab dropped down every year by 5 to 19 tonnes and the lowest catch of 23.2 tonnes were noticed during 1991-92.

A comparison of fish species composition at different time periods during the fishery within Chilka lagoon was undertaken by Biswas. Table 7.2 shows the

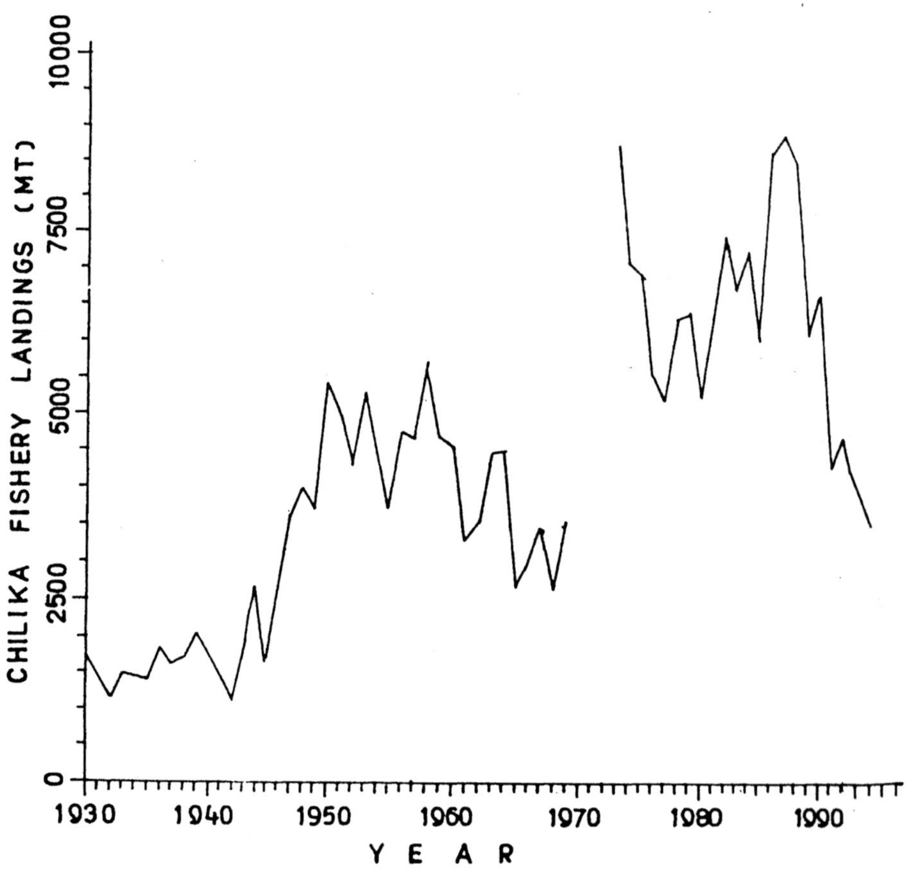

Time Series of Fishery Landings of Chilka Lagoon (Fish and Prawn Combined) Between 1930 and the Present (1993-94)

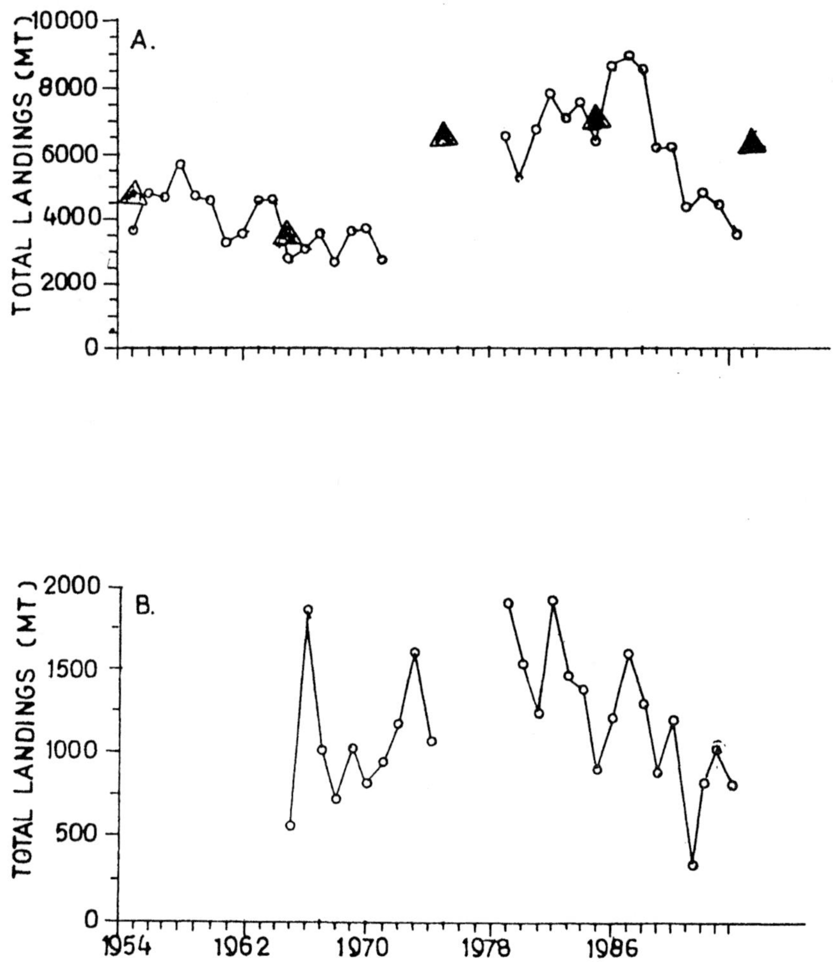

(A) History of Total Fish Landing of Chilka Lagoon (Triangles Show Decade Mean Values)

(B) History of Total Prawn Landings of Chilka Lagoon (Exclusive of Aquaculture)

estimates of the rank order of different fish groups in the commercial landings, in terms of their numerical abundance.

Table 7.2

	49/50-54/55	*57/58-65/66*	*81-91*	*91-94*
Prawns	1	1	3	2
Mullets	3	2	4	4
Clupeids	4	4	1	6
Perches	5	5	5	5
Catfishes	6	3	6	3
Threadfins	7	6	7	7
Scianids	8	7	9	8
Beloniforms	9	9	8	9
Elasmobranchs	10	10	10	–
Miscellaneous	2	8	2	1

The miscellaneous group is comprised mostly of small bodied freshwater forms, including featherbacks, eels, catfish and pearl spot. The clupeids are a diverse group of marine and anadromous species including *Anchoviella indica, Hilsa ilisa, Thrissocles kempi, Megalops cyprinoides, Elops saurus, Chanos chanos, Leignathus equalus*, and several other minor species. Over time, the relative abundance of both prawns and mullets have declined, and the relative abundance of both clupeids and miscellaneous species have increased.

A large percentage of the fish species which are exploited by the commercial fishery are migratory (either catadromous or anadromous). During the late 1950's Jhingran (1963) estimated that 63-75 per cent of the annual fish production was contributed by migratory species. More recently, this percentage has dropped to around 30 per cent due to a relative increase of freshwater species within the fishery (Biswas).

The percentage of different groups of fish and prawn, caught from Chilka lagoon from 1949 to 1993-94 in three spells i.e., 1949 to 1954, 1957 to 1965 and 1980-81 to 1993-94 are given in Table 7.3.

Prawn, which was remaining 22 to 38.7 per cent of total landings between 1949 to 1962, dropped down to 16.7 per cent in 1963 and 17.1 per cent in 1964. In 1965, however, the percentage of prawn reached the highest value of 50.8 and thereafter started falling to 17.5 in 1980-81 till the lowest value of 11.3 per cent was recorded in 1990-91. A marginal increase of 19.2 per cent in prawn catch, however, noticed in 1991-92, with further increase of 22.8 per cent in 1992-93 and thereafter fall in 1993-94 (20.6 per cent).

The percentage composition of mullet however varied between 26.9 per cent in 1957 to 10.8 per cent in 1989-90.

Clupeids, which were between 7.1 per cent in 1964 and 1965 to 16.6 per cent in 1950, showed an increase and remained between 24.7 per cent in 1980-81 to 15.1 per cent in 1990-91, and thereafter a fall of 8.6 per cent in 1991-92 and 9.1 per cent in 1993-94.

Table 7.3

Year	Fish and Prawn Groups									
	Prawn	*Mullet*	*Clupeids*	*Perches*	*Thread-fins*	*Cat fish*	*Beloni-forms*	*Scianids*	*Elasmo-branches*	*Misc. fish*
1949	26.1	14.4	15.2	8.8	4.9	9.5	1.9	5.5	0.1	13.6
1950	29.5	11.4	16.6	8.1	6.7	4.1	2.3	2.9	0.2	18.1
1951	38.7	12.6	9.8	7.2	5.3	3.2	1.9	3.0	1.7	16.6
1952	30.3	13.8	11.4	11.1	2.8	4.9	1.1	6.1	1.9	16.6
1953	22.0	15.0	14.0	13.0	2.7	10.1	0.8	5.9	2.5	14.0
1954	32.1	18.3	12.7	7.6	4.1	7.2	2.1	2.9	2.3	10.7
1957	25.1	26.9	13.5	6.2	5.8	15.5	0.5	5.9	–	0.6
1958	30.2	20.8	16.5	5.6	5.6	10.9	2.1	6.6	–	1.7
1959	31.8	14.5	13.8	6.7	9.6	10.9	2.7	4.7	–	5.2
1960	35.2	11.4	15.4	7.3	7.4	10.5	1.4	5.1	–	6.1
1961	30.2	14.2	12.3	8.3	10.8	13.3	2.8	4.1	–	3.9
1962	23.6	21.3	11.6	17.0	6.5	10.1	2.0	2.8	–	5.2
1963	16.7	25.7	12.8	15.3	4.5	12.4	0.9	8.1	–	3.6
1964	17.1	14.0	7.1	24.2	3.8	15.8	1.0	9.4	–	2.6
1965	50.8	16.9	7.1	5.9	5.3	11.9	0.7	6.2	–	3.3
80-81	17.5	19.2	24.7	11.6	4.4	7.6	2.6	–	–	12.1
81-82	24.9	17.9	19.2	6.9	4.9	6.2	1.5	–	–	18.3

(contd....)

Year	Fish and Prawn Groups									
	Prawn	Mullet	Clupeids	Perches	Thread-fins	Cat fish	Beloni-forms	Scianids	Elasmo-branches	Misc. fish
82-83	20.5	13.9	16.4	10.0	5.2	11.8	2.9	–	–	19.3
83-84	18.5	13.4	21.5	12.9	5.0	10.1	3.3	–	–	15.2
84-85	14.5	15.1	22.2	14.7	3.1	10.6	2.7	–	–	17.2
85-86	13.3	17.6	21.3	15.6	3.9	14.5	4.2	–	–	14.1
86-87	17.9	13.2	20.1	13.5	4.4	11.9	3.5	–	–	15.6
87-88	15.4	12.4	20.3	14.1	4.0	11.6	3.1	–	–	19.0
88-89	15.0	12.2	17.2	15.4	3.9	13.7	2.9	–	–	19.7
89-90	17.6	10.8	15.6	15.2	4.5	15.1	3.0	–	–	18.1
90-91	11.3	12.2	15.1	13.6	5.3	17.9	2.8	–	–	21.8
91-92	19.2	13.3	8.6	10.8	3.5	16.1	2.8	2.5	–	23.2
92-93	22.8	12.6	13.6	11.9	3.7	13.6	3.2	2.7	–	21.9
93-94	20.6	16.7	9.1	13.4	5.3	14.7	3.4	4.3	–	19.5

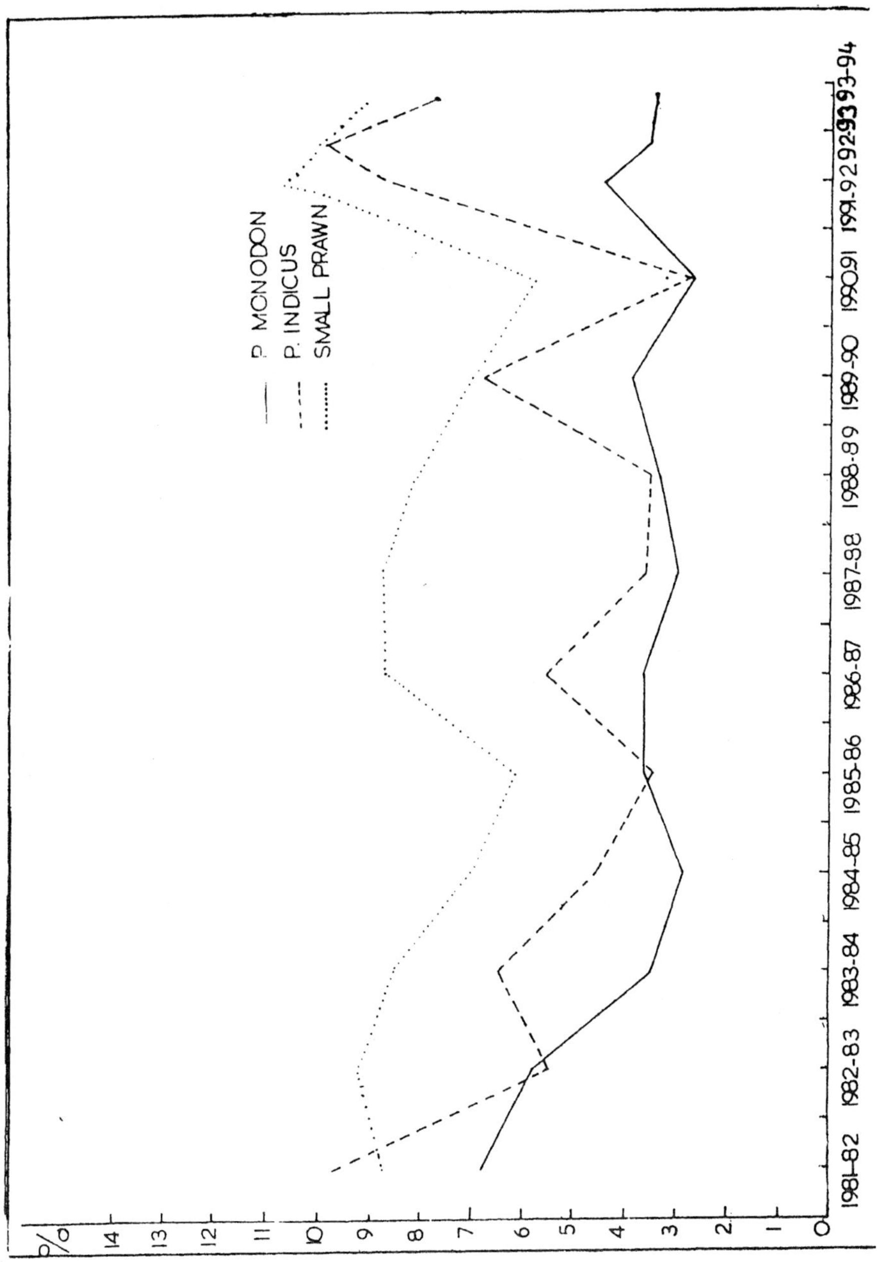

Graphs Showing Trend of Prawn Catch in Chilka Lagoon During 1981-82 to 1993-94

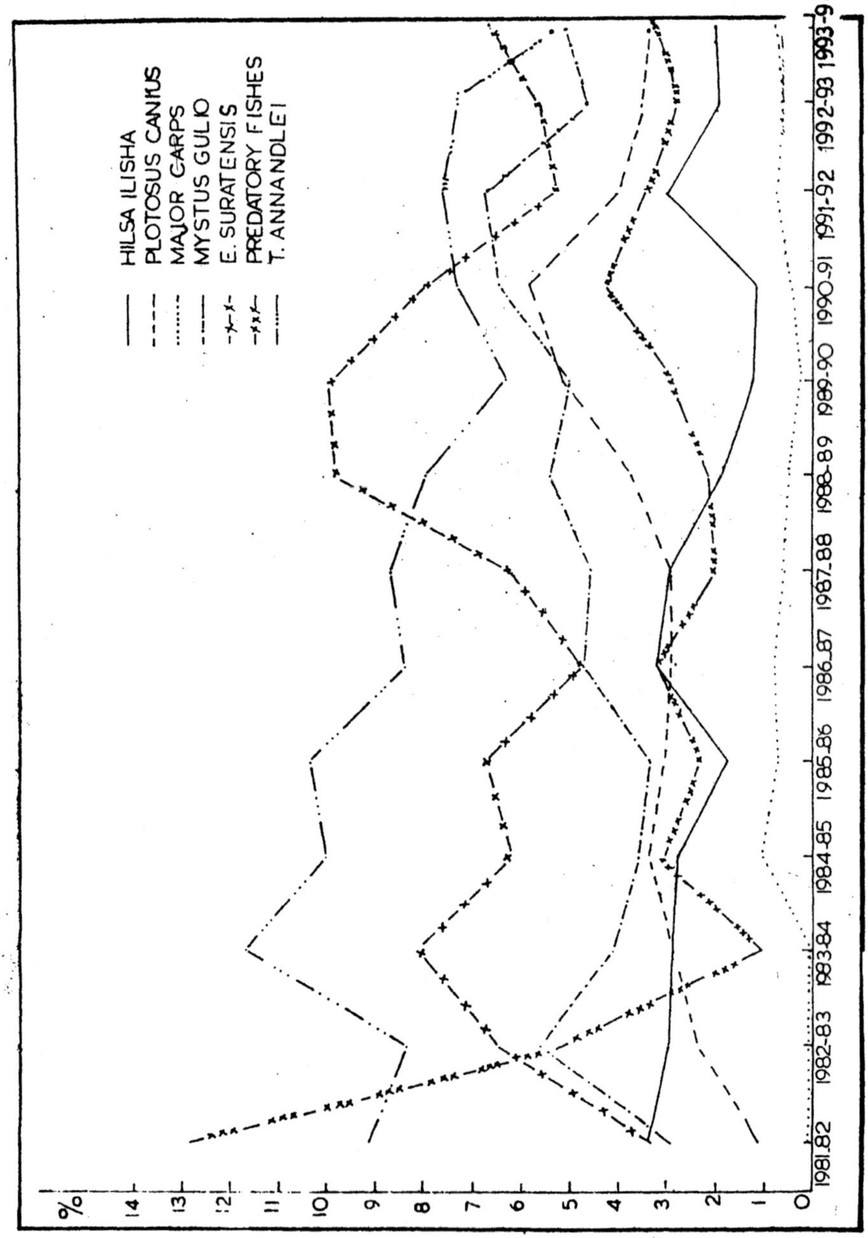

Graphs Showing Trend of Different Fish Catch in Chilka Lagoon During 1981-82 to 1993-94

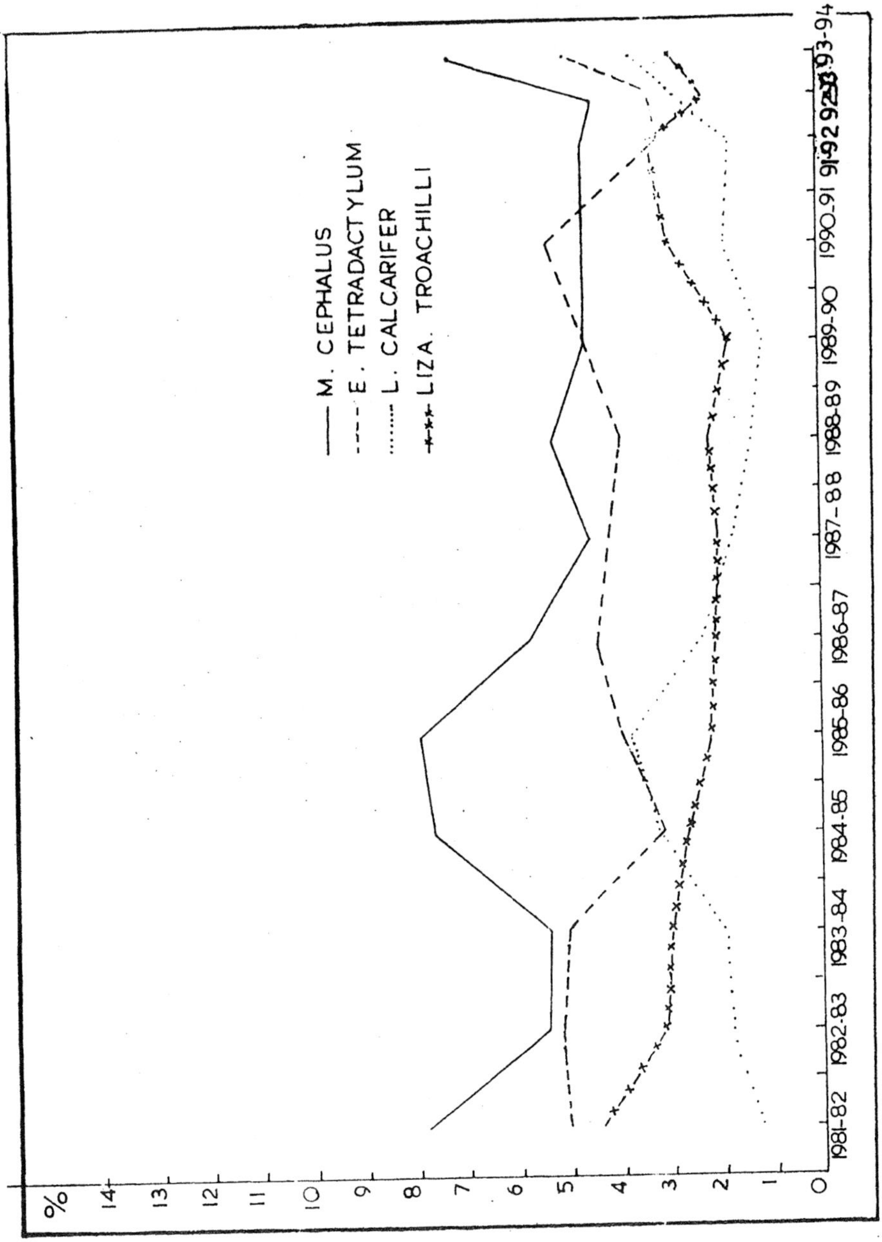

Graphs Showing Trend of Different Fish Catch in Chilka Lagoon During 1981-82 to 1993-94

Wide fluctuations in composition of perches from 5.6 per cent to 29.2 per cent is noticed between 1949 to 1982-83; and from 1983-84 to 1993-94, a variation of 12.9 per cent to 15.6 per cent is estimated.

Except in 1961, when threadfins showed highest concentration (10.8 per cent), the group remained in between 2.7 to 9.6 per cent during the entire period.

Catfish were between 3.2 per cent to 10.1 per cent during 1949 to 1954. Their population increased in 1957 (15.5 per cent) and maintained more or less similar composition except 1980-81 and 1981-82 (7.6 per cent and 6.2 per cent) and finally the highest concentration of 17.9 per cent in 1990-91.

Beloniforms, comprised of 0.7 to 4.2 per cent of the total catch during the period 1949 to 1993-94.

Miscellaneous fishes contributed 10.7 to 18.1 per cent of the catch during 1949 to1954. From 1957 to 1965, their share was 0.6 to 6.1 per cent of the total catch, which againincreasedfrom12.1 per cent in1980-81to the highest of 23.2 per cent in1991-92.

Species-wise rank order of 26 varieties of fish and four groups of prawn estimated from 1981-82 to 1993-94 is shown in Table 7.4.

Persistant upholding the higher position have been noticed in respect of the following species and groups during 1981-82 to 1993-94.

Trissocles annandalei	-	1st. position in 1983-84, 1984-85 and 1985-86.
	-	2nd position in 1986-87, 1987-88 and 1991-92.
	-	3rd position in 1981-82, 1982-83, 1990-91 and 1992-93.
	-	4th position in 1988-89.
	-	5th position in 1989-90.
	-	7th. position in 1993-94.
Etroplus suratensis	-	1st position in 1988-89, 1989-90 and 1990-91.
	-	3rd position in 1985-86.
	-	4th position in 1982-83, 1983-84, 1987-88 and 1993-94.
	-	6th position in 1984-85, 1986-87 and 1992-93.
	-	7th position in 1991-92.
	-	12th position in 1981-82.
Small prawn	-	1st position in 1986-87, 1987-88, 1991-92, 1992-93 and 1993-94.
	-	2nd position in 1982-83 and 1989-90.
	-	3rd position in 1983-84 and 1988-89.
	-	4th position in 1981-82 and 1984-85.
	-	5th position in 1985-86.
	-	6th position in 1990-91.

The species and group which once occupied first and second position during the period are—

Mugil cephalus	-	2nd position in 1984-85.
Penaeus indicus	-	2nd position in 1981-82, 1992-93 and in 1993-94.
	-	3rd position in 1989-90 and 1991-92.

Table 7.4

Species	Years												
	81-82	82-83	83-84	84-85	85-86	86-87	87-88	88-89	89-90	90-91	91-92	92-93	93-94
1	2	3	4	5	6	7	8	9	10	11	12	13	14
Mugil cephalus	5	8	7	2	2	3	7	7	11	11	8	8	3
Mugil subviridis	10	13	11	8	8	17	11	11	14	14	9	15	15
Mugil speigleri	15	21	21	24	19	21	19	25	26	24	22	22	20
Liza troschelli	8	14	15	21	21	22	20	18	18	15	16	20	19
Penaus indicus	2	7	6	7	13	4	13	14	3	17	3	2	2
Penaeus monodon	6	5	12	18	12	13	15	15	13	18	13	11	14
Small prawn	4	2	3	4	5	1	1	3	2	6	1	1	1
Chanos chanos	24	27	5	27	27	28	26	29	30	29	28	5	25
Nematolosa nasus	9	17	9	3	4	5	5	6	6	10	11	14	11
Hilsa ilisha	11	15	16	19	22	15	16	19	22	22	17	21	21
Megalops cyprine ides	17	22	26	26	26	24	27	24	23	26	24	27	27
Elops saurus	29	11	25	25	23	25	25	26	20	23	27	25	23
Trissocles annandalei	3	3	1	1	1	2	2	4	5	3	2	3	7
Lates calcarifer	19	20	19	12	11	20	23	21	24	19	20	17	13
Sparus sarba	23	26	24	22	29	27	21	20	19	21	23	23	22
Etroplus suratensis	12	4	4	6	3	6	4	1	1	1	7	6	4
Gerres setifer	25	24	22	17	13	12	18	23	25	20	21	24	24

(Contd....)

Species	Years												
	81-82	82-83	83-84	84-85	85-86	86-87	87-88	88-89	89-90	90-91	91-92	92-93	93-94
1	2	3	4	5	6	7	8	9	10	11	12	13	14
Pomodysis hosta	26	28	30	29	28	29	29	30	29	30	29	29	29
Therapon puta	27	25	29	28	24	23	24	22	21	25	25	28	28
Eleutheronema tetradactylum	7	10	8	14	10	8	10	10	12	8	14	12	8
Mystus & gulio	13	6	10	9	15	7	6	5	8	[illegible]	5	9	9
Osteogenelosis militaris	14	12	13	10	9	9	8	9	9	9	4	7	5
Plotosus canius	21	18	17	11	18	19	17	12	7	5	10	13	18
Tylosurus stronogylurus	18	16	14	20	7	14	14	16	16	16	18	16	17
Triacanthus brevirostris	22	23	20	15	17	16	12	13	15	9	12	10	12
Sciena russellii	20	19	18	15	16	10	9	8	10	12	19	19	10
Caranx carangus	28	30	27	30	30	30	30	28	28	28	30	30	30
Major carps	30	29	29	23	25	26	28	27	27	27	26	26	26
Predatory fishes	1	9	23	13	20	18	21	17	17	13	15	18	16
Misc. fish	16	1	2	5	6	11	3	2	4	2	6	4	6

Predatory fishes	-	1st position in 1981-82.
Misc. fishes	-	1st position in 1981-82.
	-	2nd position in 1983-84, 1988-89 and 1990-91.

A distinct rising trend of prawn and fish catch of Chilka lagoon from 1931 to 1937; 1942 to 1951; 1953 to 1957; 1967 to 1974; 1978 to 1985 have been noticed during last 66 years, that is, from 1929 to 1994 on the basis of their trend values (moving average of five years). Downward trend persisted for two years in 1952 to 1953; nine years in 1958 to 1966 (longest duration); three years in 1975 to 1977 two years in 1987 to 1988 and for four years in 1991 to 1994.

The fishery of the lagoon after each spell of low annual catch show upward trend which some times take as long as 9 years to recover. As such, it will perhaps not be proper to apprehend depletion of fish and prawn population of the lagoon after noticing downward trend for a few years.

The carb landings of the lagoon, however, show a distinct downward trend after initial rise from 1972 onwards till 1982 reaching its peak and thereafter fall in catch for last 11 years.

While considering the percentage of fish and prawn groups of the lagoon in three spells, prawn and mullets, the prized species of the lagoon, exhibited a downward trend, while cat fish and miscellaneous fish as a group were dominant in the annual catch in recent years when compared from their percentage composition of initial years.

On further break up of the important species and groups, the un-economic species like *Thrissocles annandalei Etroplus suratensis* and small prawns were found to occupy important position in the fish population dynamics of the lagoon in recent decades, where *Trissocles annandalei* occupied first position for three years, second position for three years, third position for four years, fourth and fifth position for one year each. *Etroplus suratensis* maintained first position for three years, third position for one year, fourth position for four years, sixth position for three years and twelfth position for one year. Small prawn uphold the first rank for five years, second rank for two years, third rank for two years, fourth rank for two years, fifth rank for one year and sixth rank for one year. Thus it can be assumed that, these above three uneconomic species and groups are gaining predominance over economic fish and crustacean population of the lagoon, keeping behind the prized fish like, *Mugil cephalus* and *P. indicus*, which once occupied second position for one and two years respectively.

Predatory fishes mostly comprising of freshwater varieties are also gaining their dominance in the lagoon by occupying the first rank within last decade and can not be ignored. Likewise, miscellaneous group, which are mostly of freshwater carps, feather backs, eel and minnows are gaining importance among the fish population of the lagoon with the changing ecological condition (freshwater conditions), since the group occupied first position once and second position twice in eighties and also in the current decade.

Prawn Fishery

Penaeid prawns are of foremost importance from the viewpoint of the fisheries of the Chilka, and the species that contribute to fishery are:

Penaeus indicus
Penaeus monodon
Metapenaeus monoceros
Metapenaeus dobsoni

Prawn constituted 22.0 to 38.7 per cent of the total catch of the lagoon during 1949-50 to 1954-55 with annual fluctuations from year to year as shown in Table7.5. The mean percentage of prawn catch during the period being 29.78.

Table 7.5

Year	*Production of prawn in tonnes*	*Total Production in tonnes*	*Percentage of prawn out of total production*
1949-50	20.96	80.31	26.1
1950-51	23.07	78.38	29.5
1951-52	32.34	83.64	38.7
1952-53	27.01	89.06	30.3
1953-54	16.10	73.76	22.0
1954-55	20.03	62.23	32.1

The prawn production from the lagoon during 1957-58 to 1965-66, though increased in volume, but their percentage in total production showed annual fluctuations with 16.72 per cent as minimum in 1963-64 and 50.84 per cent as maximum in 1965-66, the details of which are given as in Table 7.6. The mean percentage of prawn out of total catch during the period was calculated to be 28.08.

Table 7.6

Year	*Production of prawn in tonnes*	*Total Production in tonnes*	*Percentage of prawn out of total production*
1957-58	1119.48	4455.68	25.12
1958-59	1158.56	3837.88	30.19
1959-60	1209.17	3796.74	31.85
1960-61	917.79	2603.58	35.25
1961-62	863.85	2861.16	30.19
1962-63	920.62	3896.88	23.62
1963-64	657.00	3928.35	16.72
1964-65	548.24	3214.90	17.05
1965-66	1862.53	4375.63	50.84

The prawn production of the lagoon in eighties and early nineties showed a downward trend both in respect of volume and their percentage in total production with yearly fluctuations from 11.26 per cent (lowest during the period in 1990-91 to 22.99 (highest during the period in 1979-80 with an yearly average of 18.03 per cent in total catch during the period as shown in Table 7.7.

Table 7.7

Year	*Production of prawn in tonnes*	*Total Production in tonnes*	*Percentage of prawn out of total production*
1979-80	1550	5169	22.99
1980-81	1101	6287	17.50
1981-82	1857	7466	24.87
1982-83	1373	6696	20.51
1983-84	1339	7250	18.47
1984-85	871	5999	14.51
1985-86	1142	8589	13.30
1986-87	1589	8872	17.91
1987-88	1304	8475	15.39
1988-89	917	6096	15.04
1989-90	1178	6670	17.65
1990-91	481	4273	11.26
1991-92	840	4545	19.2
1992-93	947	4159	22.8
1993-94	687	3329	20.6

The fishery of *Penaeus indicus* in Chilka lagoon is chiefly attributed to two juvenile waves designated as April-July wave and August-December wave, of which April-July wave is more important (43.08-64.9 per cent). The success of the fishery depend on the successful wave, which may start late (May) or early and sooner (June) or later (August) constituting a detection size of 93-101mm. and departure size of 101-109 mm. For the success of the wave, post-larval abundance in February (January-March) is required.

August-December wave plays a secondary role in yield (10.48-45.17 per cent),which in certain years commences early (July) or late (September) with an overall average size of 84-91 mm. and departure size of 108-124 mm. Post larval abundance in May (April-July) is attributed for the success of the wave.

For success of all juvenile waves mentioned above, post-larval abundance commencing in October and lasting till July of the following year with three peaks are indicated.

Penaeus indicus feed on detritus as dominant item of food, followed by plant matter, molluscs, crustaceans remains, algae, diatoms and sand; and grow at the rate of 36 mm. per month till they attain 110 to 140 mm. size at capture.

The fishery of *Penaeus monodon* show extreme fluctuations, good catches are usually made between March and June, but may commence in February and extend to July. In certain years appreciable catches were also reported during September and November period.

February-July wave is by far more dominant and accounts for nearly the entire fishery. The post larval abundance during the period October-January has a bearing on the success of this wave. The August-December wave plays a very minor role in its contribution to the fishery, the success of which depend on the post-larval abundance in April-May.

The mean monthly growth of the species of February-July wave varies from 9.5 mm. to 12.4 mm. with a maximum growth of 205 mm. in November. The average monthly growth of August-December wave register 12.2 to 16.3 mm. in different regions of the lagoon, attaining 180-225 mm. in February-March of the following year.

Factors Influencing Prawn Fishery

Coastal current, wind direction and velocity, coastal movement of prawns as well as survival of post-larvae from planktonic to benthic life and mortalities of early juveniles from predation are all factors which can, individually or cumulatively affect the fishery of the lagoon.

The strategic zone, occupying the pathway between the main lagoon and the outer channel covering prawn fisheries e.g. Godhi, Bubalia, Baruna, Diankudi, Maraja, Kholamuha, Ganserra and Kathapenta, situated a little south-west to Magarmukh facing the lagoon ranks first in prawn productivity. The main prawn season of the area is March-May or June depending upon rains.

The zone including the eastern shore of northern sector from 42-Bahar Chilka to Haridas near Daya river mouth with important prawn grounds like 42-Bahar Chilka, Haridas and Karimpur ranks second in productivity extending prawn season from March to July.

The zone covering at the commencement of the outer channel and encompasses Berhampur, Mohosa and southern creeks upto Baliarkhati and on the west covering the turn of outer channel including Ratamati occupies third place-in prawn productivity. The Chief prawn season for this zone is March to June, but in creeks after rains in July and August.

Southern portion of the lake including Palur Madurchua, Gajapatinagar, Kumarpur, Rambanona and Sabulia ranks fourth in prawn production throughout the most part of the year.

Prawn season of areas from Sankuda and Samala to Balugaon extend variously from March-May and July to October ranks fifth in prawn production.

The zone extending from Borago partly into southern sector ranks sixth in prawn production after rains.

The areas from Nairi to Mangalajodi (neat Kaluparaghat) register lowest prawn catch and the prawn season of the area extends from March-May and from August-October.

Crab Fishery

The crab fishery, chiefly represented by the lone species *Scylla serrata*, has fairly wide distribution in the lagoon, common in Rambha zone, Balugaon-Nalabana Kathapenthaghat zone and Baradihi-Magarmukh zone, where average annual salinity varies from 5-14 ppt. Juveniles of the species are available from middle of January to middle of April and peak fishing season extends from middle of September to last part of December, Fishing ground of the species is located at Malud, Parikud, Khallikote and Rambha. *Neptunus pelagicus*, next important to *Scylla serrata* are found in outer channel rather than the lagoon, where average annual salinity range between 16.5 to 22.5 ppt. Less catch is obtained from Rambha zone. Good catch from June-January with peak in August to October is obtained.

Main fishing grounds are Satapada, Nuapada and Alupatna, Balugaon-Nalabana-Kathapenthaghat and Baradihi Magarmukh zone contribute 77.23 per cent of total crab landings on an average. Rambha zone gives 19.84 per cent of total crab landing on an average, while 2.90 per cent of average landing is obtained from Barakudi-Bhusandapur zone.

Crab landings from the lagoon from 1980-81 till 1993-94 are given below.

Year	*Crab landing in tonnes*
1980-81	122
1981-82	80
1982-83	88
1983-84	141
1984-85	90
1985-86	79
1986-87	54
1987-88	39
1988-89	44
1989-90	36
1990-91	24
1991-92	30
1992-93	15
1993-94	12

A progressive downward trend in crab landings in Chilka has been noticed in eighties and early ninetees, which again warrent the depletion of this economic species, if suitable measures are not been taken in time.

Monthly Landings of Fish and Prawn

From the monthly landings of major fish species and groups in recent years (1990-91 to 93-94), it appears that catch of *Mugil cephalus* in 1990-91 were predominant between April and July (10.7 to 14.2 per cent) of total yearly catch; while in 1991-92, 92-93 & 93-94 the catch of the species was more between September to December than the rest of the year, when the salinity of Central zone was between 5 to 5.7 ppt. and of Southern zone between 8.2 to 8.7 ppt.; where the species dominated in the catch. Other mullets namely *Mugil subviridis*, *Mugil spegleri* and *Liza troschelli* also exhibited higher landings between September to January in the year 1991-92 (9.8 to 11.3 per cent), 1992-93 (12.0 to 12.7 per cent), and 1993-94 (10.2 to 11.6 per cent); except in the year 1990-91; when 12.1 per cent catch of other mullets were obtained in the month of July.

Table 7.8

Mullets	*Mugil cephalus* per cent of catch in different months and years				*Other mullets* per cent of catch in different months and years			
	90-91	*91-92*	*92-93*	*93-94*	*90-91*	*91-92*	*92-93*	*93-94*
April	11.8	3.8	7.3	6.8	8.3	6.9	10.5	6.2
May	11.0	3.9	8.2	9.4	8.1	7.4	11.7	7.4
June	10.7	6.0	6.0	6.3	7.2	6.7	6.1	5.0
July	14.2	4.5	8.2	5.7	12.1	6.4	10.3	5.3
August	7.8	4.4	7.0	8.5	8.2	5.7	7.9	7.7
September	5.5	4.4	14.5	9.9	6.5	6.3	12.7	9.0
October	5.8	7.4	14.1	10.8	6.9	10.6	12.0	11.1
November	6.4	6.8	6.9	10.1	7.6	9.8	7.4	10.2
December	6.3	6.7	4.8	9.7	8.8	10.7	5.1	11.6
January	6.5	6.5	7.6	8.4	7.9	11.3	6.7	10.9
February	5.9	5.1	7.3	6.7	7.4	9.4	6.5	7.7
March	7.9	6.9	8.6	7.6	10.8	8.8	5.8	7.8
Total catch in tonnes	195.29	328.19	205.93	245.64	328.39	388.23	318.89	310.91

Higher monthly catch of all varieties of prawn namely *P. monodon* (tiger); *P. indicus* (white) and small prawns have been observed between April to August in all the 4 years i.e. from 90-91 to 93-94 with annual variation as shown in Table 7.9.

During this period i.e. between April to August the average salinity in different zones are higher than the rest of the months of the year indicating the catch of more prawns in the lagoon due to optimum environmental condition thus prevail with respect of higher salinity gradient in different sectors of the lagoon.

Several fish species, such as, *Etroplus suratensis* and *T. annandalei* are establishing breeding population within the lagoon and are capable of completing the life cycle within the brackishwater environment of Chilka, having different compatible ecology and salinity gradient.

This can be observed from the higher catch (8.2 to 14.8 per cent) of *E. suratensis* between April to July during 90-91 to 93-94 when the salinity of lagoon water in all the zones are highest during the year.

T. annandalei however, exhibited the highest landings (8.8 to 14.6 per cent) between January to May in all these years indicating the establishment of the species to different salinity gradient of the lagoon. (Table 7.10)

Table 7.9

Months	*P. monodon (Tiger)* per cent of catch in different months & years				*P. indicus (white)* per cent of catch in different months & years				*Small prawns* per cent of catch in different months & years			
	90-91	*91-92*	*92-93*	*93-94*	*90-91*	*92-92*	*92-93*	*93-94*	*90-91*	*91-92*	*92-93*	*93-94*
April	21.1	17.2	20.4	6.3	20.9	4.3	19.5	4.7	25.1	8.8	17.4	9.0
May	19.6	9.7	9.5	8.4	18.4	12.4	18.3	6.2	16.5	10.9	15.5	10.6
June	10.4	11.2	9.2	10.1	9.1	23.3	6.7	22.1	8.1	16.3	8.5	15.6
July	7.6	12.8	16.2	16.8	8.2	19.8	13.4	39.0	11.9	16.3	16.1	18.3
August	7.9	9.8	19.3	13.1	9.0	12.7	29.1	11.1	9.8	8.9	17.0	15.4
September	4.5	7.9	10.6	7.7	7.9	9.0	5.9	5.2	5.0	7.0	7.4	6.4
October	3.5	8.7	8.4	7.9	4.3	5.9	3.4	3.8	3.2	4.3	4.2	5.9
November	4.7	4.3	2.1	6.7	4.4	2.3	0.8	2.3	2.5	1.2	1.7	4.4
December	6.9	3.0	0.9	9.8	3.4	1.3	0.3	2.3	2.7	1.5	1.5	4.3
January	2.5	2.9	1.3	4.9	2.7	1.6	0.7	1.5	1.0	3.7	2.7	3.2
February	2.3	2.5	1.8	2.8	2.7	4.2	0.8	0.7	2.5	11.2	4.8	2.9
March	8.2	9.8	3.0	5.5	8.6	3.2	1.0	1.1	11.5	9.8	3.1	3.8
Total catch in tonnes	113.35	162.20	153.85	121.69	120.62	285.55	380.45	261.29	255.10	428.39	413.26	304.34

Table 7.10

Months	*E. suratensis* per cent of catch in different months & years				*T. annandalei* per cent of catch in different months & years			
	90-91	*91-92*	*92-93*	*93-94*	*90-91*	*91-92*	*92-93*	*93-94*
April	14.8	9.6	7.9	11.4	10.0	6.9	12.4	9.9
May	11.9	9.9	7.3	12.9	8.4	6.4	12.0	7.0
June	8.2	9.4	11.2	5.3	8.1	6.1	9.0	4.4
July	11.7	8.5	10.7	7.5	9.8	5.7	10.2	5.2
August	8.0	6.3	6.7	9.3	6.9	5.5	7.7	6.3
September	5.9	6.2	7.8	8.5	5.4	5.0	7.7	6.9
October	6.1	8.7	10.4	7.9	6.6	6.5	8.1	7.2
November	6.7	6.8	7.5	7.4	7.0	10.5	8.8	6.9
December	6.5	8.3	5.4	8.4	8.4	9.9	4.7	8.2
January	6.2	9.9	8.3	8.5	9.5	9.4	5.7	14.6
February	6.4	7.1	7.8	5.6	8.8	12.0	6.9	11.4
March	7.5	9.2	8.9	7.2	11.1	15.8	6.5	11.9
Total catch in tonnes	339.08	239.59	241.42	224.42	309.86	427.81	300.63	179.95

The miscellaneous groups mostly of featherbacks, eels, cat fish have been found to be dominant in the catch between April to July. *Mystus gulio* were caught in large numbers (8.5 per cent to 19.7 per cent) between March to June in all these 4 years (90-91 to 93-94), when high salinity condition prevails in the lagoon than rest of the year. Freshwater fish comprising of featherback, murrel, cat fish, eel are landed in more quantities (9.4 to 14.6 per cent) between February to March during 90-91 to 92-93 mostly in the Northern sector of the lagoon; where the average salinity between 0.7 to 1.2 ppt. prevails in those months. In 1993-94 however, the highest catch of 11.3 per cent was obtained in the month of May.

The occurrence of miscellaneous fish in large quantities was reported between April to July during 90-91 (9.1 to 13.5 per cent); between January to March in the year 91-92 (10 to 11.7 per cent); between August to October in 92-93 (11.6 to 12.2 per cent) and during March in 93-94 (10.5 per cent) indicating that these miscellaneous groups have established into the different sectors of the lagoon irrespective of monthly salinity fluctuations prevail in the lagoon during monsoon months (August to December) and the average salinity dropped between 0.2 to 0.5 ppt. in the Northern sector; 5.2 to 5.7 ppt. in central sector; 8.2 to 8.7 ppt. in Southern sectors and 0.6 to 6.8 ppt. in connecting channel to Bay of Bengal. (Table 7.11)

A large percentage of fish species in Chilka lagoon can be classified as consumers feed directly on plant materials and decaying detritus. Because of this, trophic characteristic fish population of Chilka lagoon are extremely productive

Table 7.11

Months	*M. gulio* per cent of catch in different months & years				*Misc. fishes* per cent of catch in different months & years				*Freshwater fishes* per cent of catch in different months & years			
	90-91	*91-92*	*92-93*	*93-94*	*90-91*	*91-92*	*92-93*	*93-94*	*90-91*	*91-92*	*92-93*	*93-94*
April	11.4	10.8	10.9	6.6	12.1	6.6	10.2	7.3	10.6	10.7	11.6	8.6
May	19.7	10.5	11.8	15.0	11.7	7.2	6.4	8.7	9.6	11.0	10.0	11.3
June	8.9	10.6	10.4	8.5	9.1	7.9	8.7	8.3	6.7	10.9	9.0	7.9
July	10.5	11.9	11.9	7.5	13.5	7.4	9.8	6.9	7.7	10.1	10.8	7.7
August	7.2	6.7	7.2	6.9	8.4	6.1	11.6	9.1	6.8	5.8	7.3	6.8
September	4.5	7.4	8.1	10.0	5.7	5.9	11.8	8.0	6.4	5.6	8.1	7.6
October	4.7	8.2	9.5	9.4	5.6	8.5	12.2	8.7	7.6	5.9	9.1	7.5
November	5.1	6.1	5.5	6.6	6.4	7.8	7.8	7.3	7.7	6.1	8.0	7.5
December	5.0	5.8	4.7	9.2	6.3	9.8	3.7	10.3	7.6	7.3	4.8	8.1
January	4.8	9.8	6.8	7.4	6.9	10.8	6.1	6.5	8.3	8.6	6.2	9.0
February	6.4	4.9	6.3	6.3	6.1	10.0	5.2	8.2	9.4	7.4	6.5	7.8
March	11.7	7.0	7.0	6.4	8.0	11.7	6.4	10.5	11.2	10.6	14.6	9.9
Total catch in tonnes	274.70	278.46	204.16	175.66	316.60	263.93	298.58	190.75	205.33	188.78	148.16	145.71

(quantified as grams carbon placed on an areal basis) in comparison to fish population in other aquatic environments. Most fish species recruit to the fishery in relatively small average size (and low mean age) indicating the possibility of growth over fishery and an inefficient fishery utilisation of the resources.

It has been estimated that the annual fish yield from the lagoon averaged about 90 Kg. per ha. during 1980s (ranged being 65 to 122 Kg. per ha.)

8

Role of Cooperatives

Central Fishermen Co-operative Marketing Society Ltd. (CFCMS Ltd.) an Apex society, came into existence in July 1959, basing on the recommendations of Mr. Leid Law. Every year the Revenue Department presents a list, with all the fisheries sources with an annual demand for lease-payment to the Apex society. The lease value was paid in four instalments annually. The Apex subleases the same sources to the primary fishermen co-operative societies (PFCS) in the name of "Leader bahania" (Head fishermen).

Some fishery sources are not leased out through the Apex society. Some Uthapani, some smaller jano fisheries and chulimunda are leased out directly by the Tahasildars. (Representative of Revenue Deptt.).

After the formation of Central Fishermen cooperative Marketing Society Ltd. (CFCMS Ltd.) the following primary Fishermen Cooperative Societies have been organised in Chilka areas for fishing in 1960-61 as shown in Table 8.1.

Table 8.1

S. No.	*Name of the society*	*Villages which constitute the society or area of operation*	*No. of members*	*Fisheries assigned to the society*	*Licence fee of the fisheries (1960-61) N.B. in every year 10% increases (Rs.)*
1	*2*	*3*	*4*	*5*	*6*
1.	Satpada Primary Fishermen Cooperative Society	Satpada	330	Panchubatia-Chandbatia	207
				Bahar Munda—	75
				Sida Nadi-Niataghar	383
				Sarkhanda,	882
				Barpania	2518
					495
2.	Kalupada Primary Fishermen Cooperative Society	Dadhibabanpur	66	Gorada	916
		Bali Nasi		Jagadala	16
		Panchupatia		Golara	9
		Panidwar		Haridas	117
3.	Berhampur Primary	Berhampur	312	Marai-Jano Bahana	5317
				Sukrapada	716

(contd....)

1	2	3	4	5	6
	Fishermen			Rambhartia	1410
	Cooperative			Chabiskhanda	1797
	Society				
4.	Kanakdurga	Khiri Sahi	279	Khirisahi	179
	Primary	Biripadar		Dauli	440
	Fishermen	Baliada		Cinapalli	1870
	Cooperative	Singhanandi		Akesamada	147
	Society			Kansari Nadi	734
				Hati Khala	119
				Khandia Nadi	887
5.	Ambika Primary	Chandrapat	482	Solakuda Chhadra	273
	Fishermen	Nuapada		Kandakhar	3300
	Cooperative				
	Society				
6.	Durga Primary	Bhimpur	239	Nalabana	15694
	Fishermen	Banpur		Barunikhati	4149
	Cooperative	Bidharapur			
	Society	Galua			
7.	Jogomaya	Mudirath	188	Gopata	615
	Primary	Karimpur		Chandramakhala	110
	Kautakudi	Kautakudi		Panibata	26
	Cooperative			Talagadagadia	99
	Society			Kimbhiria	514
				Karandia	3575
				Jamuna Kudi	275
				Baisibati	789
				Nuagaon	770
				Baisibatibundh	96
				Mudirath Granthala	93
				Bandha Singitali	42
				Bahar Chilka	119
8.	Jagulei Primary	Barunapada	100	Khandi river	9
	Fishermen	Todapada		Barpania(non-prawn)	1478
	Cooperative	Matiapada		Nuakhar Majhi Adang	116
	Society	Tangi		Dharanikuda	176
		Chandpur		Khasmahal portions	1690
				Basantapur	205
9.	Gopinathpur	Gopinath-	304	Bhobania	8360
	Primary	pur		Kantai Bhodasipala	3927
	Fishermen	Jadupur		Panda Nali	338
	Cooperative			Upara Badagada	587
	Society			Raipur Nali	339
				Kanta Kudi	312
				Daisikhia Muhana	253
				Sundi Jharia	1160
10.	Mangala Primary	Baghanagi	101	Kodinbati	4180
	Fishermen	Barabakudi		Bharanikudi	
	Cooperative			(Anchal portion)	1890
	Society				

(contd....)

1	2	3	4	5	6
11.	Lakshmi Primary	Nairi	230	Bhandisal	1100
	Fishermen	Baulabandha		Dharuoraguahasi	1692
	Cooperative	Kamandal		Khanimunha	3300
	Society	Santrapur		Sankha	2567
				Kaluri Hadi	125
				Dharaora Quachupi	37
				Sankha Hadi	83
12.	Narayani Primary	Santrapur	558	Tantadhua	248
	Fishermen	Badakul		Hauli	146
	Cooperative	Langaleswar		Godabata	110
	Society	Pathara		Gorasara Kahanda	1155
				Chilka	252
				Sagakhia Mahunia	619
				Upara Chakra	825
				Kurupala	303
				Tarakasi	138
				Bhuruti	706
				Madhurapani	294
				Panohakuda	624
				Talachatara	596
				Jayanala	1295
				Kudakudi	1443
				Panda	286
				Barunakhati	12,525
				Panibata	2200
				Baliada Khati	587
				Ghikuda & Khdiakuda	352
				Oazasara	917
				Khanimuha Uchali	664
				Banamali Talakhamar	
				Khanda	405
				Kholanasi	224
				Gayanala	954
13.	Jageswari	Gopinathpur	289	Chutia Nai	12
	Primary			Bahar Kanheipur Nala	20
	Fishermen			Ratamati	645
	Cooperative			Dandihira	173
	Society				
14.	Lankeswari	Alanda	110	Ambritandhi	565
	Alanda Patna	Patna		Kankajagada (Gramapali)	619
	Primary	Mahisa		Ostapalli	385
	Fishermen			Solakudi-Tala	
	Cooperative			Bhitara Chingudi	208
	Society			Mahis Gramatali	
				Bahani	
15.	Kasapur Primary	Kesapur	374	Machakhada	220
	Fishermen	Tantuliapada		Diankudi Kana	
	Cooperative	Kaluapadar		Balia Chakra	52
	Society	Sambla		Chikalakhani	

(contd....)

1	2	3	4	5	6
		Matha Sambal		Jali Muha	32
				Badasanka	25
				Jodajano	241
				Kala Kudi	400
				Kalaghar Nadi	133
				Deep Chilka	182
				Budhi Bara	
				Petha-North & South	1642
16.	Trinath Deb Primary Fishermen Cooperative Society	Rasakudi	347	Gadia	281
		Khaiakudi		Talaribundh	550
		Kholamuha		Tamanda Bana	46
		Balinasi		Matia	1485
		Malaswari		Madhara	110
		Sahabajapur		Sialachakra	660
		Naba		Baligutha	510
				Jaliguha	
				Jalimuha	1335
				Biluanari	
				Khalormuha	358
				Khatiakudi	663
				Rasakudi	377
				Gobari	2860
				Hata	743
				Baghada	305
				Sundhi Nala	48
				Baliachatara	112
				Sialabandha	77
				Badachingudi	
				Budhiakuda	29
				Gohir Khanda	
				Uthanikuda	181
				Marinalakhanda	
				Kothanikasi	
				Bahana	170
				Jalimuha Baluanasi	
				Uthapani Khati Khanda	
				Rasakudi	44
17.	Rambha Primary Fishermen Cooperative Society	Rambaha	784	Kalajora	
		Sabilia		Faraka Nasi	294
		Madhurchua		Jhara Nadi	743
		Bauria Patna		Kalijhara	31
		Tikarpada			
		Laxmipur			
18.	Haimabati Primary Fishermen Cooperative Society	Sambala	239	A portion of deep Chilka	44
		Sabilia			

(contd....)

1	2	3	4	5	6
19.	Chamunda Primary Fishermen Cooperative Society	Kumarpur Ramalanka Patanasi	347	Badakuda	169
				Maduabaharpara	88
				Badachara	16
				Syamagali Jano	68
				Siradu Jano	65
				Podakhia	3594
				Gouada Gopalia	1430
				Kumarpur Bahar Chilka	66
				Tala Palli Tala	30
				Chikili Chirgudi	
				Manda Kaduachar	107
20.	Ramachandi Primary Fishermen Cooperative Society	Damodarapur		Palur Canal	290
				Kansari Ganda	
				Khari Nala	361
21.	Chilka Primary Fishermen Cooperative Society	Paniduar Digiri Sahi Jagarnathpur Patna	116		

The above fishery societies were given fishery sources mostly for Jano and for prawn fishing. Besides in the same year another 13 Primary Fishermen Cooperative societies have been organised in Chilka area for net fishing (Bahani) in areas near by their village as shown in Table 8.2.

Table 8.2

	Name of the Society	*Village*	*No. of members*
1.	Gangadebi Primary Pathara Fishermen Cooperative Society	272	
2.	Chandi Primary Fishermen Cooperative Society	Kainchpur	231
3.	Bansiswan Primary Fishermen Cooperative Society	Khallikote	226
4.	Chandi Devi Primary Fishermen Cooperative Society	Bankiapalli	116
5.	Bhagabat Primary Fishermen Cooperative Society	Rambha	76
6.	Nairi Primary Fishermen Cooperative Society	Nairi	667
7.	Uttar Chilka Primary Fishermen Cooperative Society	Jagulipadar Kaundakhani Gadisagoda	1066

(contd....)

	Name of the Society	Village	No. of members
8.	Soren Primary Fishermen Cooperative Society	Soran	1094
9.	Bhusandpur P.F.C.S.	Bhusandapur	320
10.	Mangalajodi P.F.C.S.	Mangalajodi	235
11.	Gangakali P.F.C.S.	Baulabandha	106
12.	Bhagabati P.F.C.S.	Banapur	468
13.	Boradi P.F.C.S.	Barudi	363

Besides allotting fishery sources to the Primary Fishermen Cooperative Society, individuals (Leader fishermen) have been allotted the following fishery sources to operate as detailed below— .

Gaurang Behera	Ginapalli	626
Khetra Behera	Bhusandpur Nadi	194
Badaora Sanaora etc	Subhadrapur River	1736

Subsequently another 34 Primary Fishermen Coop. Societies were formed in the Chilka areas till 31.3.91 with the following members (Table 8.3).

Table 8.3

Sl. No.	Name of the P.F.C.S.	Whether working or not	No. of members
1.	Uttarayani P.F.C.S. (Boradi)	Liquidated	17
2.	Kuhudi P.F.C.S.	Liquidated	87
3.	Jaradevi P.F.C.S.	Working	84
4.	Refusee P.F.C.S.	Defunct	53
5.	Utkal Kamala P.F.C.S.	Defunct	33
6.	Mamu Bhanja P.F.C.S.	Liquidated	11
7.	Balugaon F.C.S	Moribond	271
8.	Uttarayani P.F.C.S. (Balugaon)	Working	62
9.	Mangala P.F.C.S.	Moribond	217
10.	Arakhakuda P.F.C.S.	Working	402
11.	Jadapur P.F.C.S.	Working	79
12.	Bagadevi P.F.C.S.	Working	126
13.	Maa Palengeswari P.F.C.S.	Working	32
14.	Budhima P.F.C.S.	Working	100
15.	Chauderdev P.F.C.S.	Working	115
16.	Barunei P.F.C.S.	Working	239
17.	Harichandi P.F.C.S.	Working	78
18.	Parbatipur F.C.S.	New 90-91	31
19.	Padmadevi P.F.C.S. (At-Jaganathpur)	Working	195
20.	Bada-Bahania P.F.C.S (At-Balipatapur)	Moribond since 4 years	370

(contd....)

Sl. No.	*Name of the P.F.C.S.*	*Whether working or not*	*No. of members*
21.	Basadeva P.F.C.S. (At-Dhanaghar Sahi)	Working	139
22.	Ugratara P.F.C.S. (At-Kantalbai)	Moribond since 4 years	50
23.	Chilka Decereesahi P.F.C.S. (At-Decree Sahi)	Moribond since 4 years	183
24.	Chilka-Jaganathpur P.F.C.S (At-Jaganathpur Patna)	Moribond since 4 years	195
25.	Kalupada (At-Kalupadaghat)	Liquidated	27
26.	Mahaveer P.F.C.S.	Moribond since 4 years	74
27.	Maa-Bhargavi P.F.C.S. (At-Kalupadaghat)	Moribond since 4 years	290
28.	Baseli P.F.C.S.	(New)	22
29.	Gajapatinagar P.F.C.S. (At-Gajapatinagar)	Moribond	519
30.	Sabalia P.F.C.S. (At-Sabalia)	Working	84
31.	Pathara P.F.C.S (At-Pathara)	Working	405
32.	Khatamunha P.F.C.S.	Working	53
33.	Jemamani P.F.C.S.	Working	51
34.	Mahamaya P.F.C.S.	Working	230

These 68 Primary Fishermen Cooperative Society with 16,117 members were formed for exploitation of fisheries in Chilka lagoon and existed till 31.3.91. Out of these 68 P.F.C.S. 49 societies availed lease of fishery sources of Chilka lake from 1988-89 to 1990-91 as shown in Table 8.4.

Table 8.4

Sl. No	*Name of the P.F.C.S.*		*Name of the Source Allotted*
1.	Berhampur P.F.C.S.	1.	Marei Jano & Bahani
		2.	Kelikuda Jano
		3.	Rambheratia Prawn
		4.	Chabisakhanda Prawn
		5.	Sukurapada Prawn
2.	Jageswari P.F.C.S.	6.	Chutianalo Jano
		7.	Bhaharkanheipur Nali Jano
		8.	Dandihida Jano
		9.	Badanalo Uthapani
		10.	Ratamati Prawn
3.	Rambha P.F.C.S.	11.	Podakhia Jano
		12.	Mundachar Kanguachar Prawn
		13.	Jhar Jano Chedara Jano of Prayagi
		14.	Chilika Fisheries Madhurachua Prawn
		15.	Chilika Fisheries Rambha, Plot No. 15, Prawn

(contd....)

Sl. No	*Name of the P.F.C.S.*	*Name of the Source Allotted*	
		16.	Kalajhar Nadi at Sabalia
4.	Chamunda P.F.C.S.	17.	Bahar Chilika Bahani, Ramalenka
		18.	Garad Gopalia Jano
		19.	Kumarpur Baharchilika Prawn
		20.	Kaduabaharfer Jano
		21.	Matiakhal Dian
		22.	Bahar Chilika Bahani, Patanasahi
5.	Harachandi P.F.C.S.	23.	Kanakuturi Jano
		24.	Jagadahana Jano
		25.	Boraga Jano
		26.	Baharachilika Bahani, Nuapada
		27.	Gaibate Prawn
6.	Satapada P.F.C.S.	28.	Panchabatia & Chandabatia
		29.	Bahar Adanga Jano
		30.	Seduanadi Nayatghar Prawn
		31.	Badamarada Prawn
		32.	Serakhanda Prawn
		33.	Nuakhar Majhi Adanga Prawn
		34.	Panchudera Prawn
		35.	Barapania Prawn
7.	Mahisa P.F.C.S.	36.	Mahisa-gramatali Bahani
		37.	Ostapalli Prawn
8.	Sabalia P.F.C.S.	38.	Deep Chilka
		39.	Kalajnara Nadi
9.	Mahamaya P.F.C.S.	40.	Chikili Jano
		41.	Garasar Pahanda Jano
		42.	Panda Jano
		43.	Jayamalla Jano
		44.	Panchakot Jano
		45.	Kalajhar Nadi at Langaleswar
		46.	Barunkhati Prawn 25% (from 16th April)
		47.	Gheemuda Prawn 25%
		48.	Jayamalla Prawn
		49.	Gerasar Pahanua Prawn
10.	Baradi P.F.C.S.	50.	Bahar Chilka Bahani, Baradi
11.	Jenamani P.F.C.S.	51.	Ghodabat Jano
		52.	Tenadhua Jano
		53.	Barunkhati Prawn 15% (From 16 April)
		54.	Gheemuda Prawn 25%
		55.	Kudakudi Jano
12.	Alandapatna P.F.C.S	56.	Amurtanidhi Jano
		57.	Sagakhia Kahunia Jano
		58.	Kankadagada Jano, Bahani, Gramatali
		59.	Chadheyanasi Chingudi
		60.	Bali Anla Chingudi
		61.	Tininalia Dian
13.	Soran Chilika P.F.C.S.	62.	Bahar Chilika Bahani, Soran

(contd....)

Sl. No	Name of the P.F.C.S.		Name of the Source Allotted
14.	Gangadevi P.F.C.S	63.	Baharchilika Bahani, Pathara Uparsahi.
15.	Chaubaradev P.F.C.S.	64.	Chandanamara Nadi Bahani
16.	Bhusandapur P.F.C.S.	65.	Baharchilika Bahani, Bhusandapur & Balipatapur
		66.	Bhusandapur Nadi Jano
17.	Kespur P.F.C.S.	67.	Machhakheda Jano
		68.	Kalajhar Farakanasi Jano
		69.	Chikalakhani Jalimuhan Jano
		70.	Diankudikona Baliachatra
		71.	Budhibar pantha (South) Chingadi
		72.	Budhibar Pantha (North)
18.	Budhima P.F.C.S.	73.	Manikpatna Nadi Bahani, Gangadharpur
		74.	Dandihidamuhan Prawn
		75.	Gangadharpur Nadi Bahani
		76.	Nadiakhia Jano
19.	Padmadevi P.F.C.S.	77.	Baharchilika Bahani, Bhusandapur & Balipatpur
20.	Pathara P.F.C.S.	78.	Baharchilika Bahani, Pathara Talasahi
21.	Durga P.F.C.S.	79.	Ghodabata Prawn
		80.	Boraganasi Prawn
		81.	Kathapentha Prawn
		82.	Gahaneswarnasi Prawn
		83.	Barunkhati Prawn
		84.	Kouliboraga Uthapani Prawn
		85.	Jagadhana Uthapani Prawn
22.	Kanak Durga P.F.C.S.	86.	Garagari Jano
		87.	Ginapali Jano
		88.	Deuli Jano
		89.	Khirichi Jano & Bahani
		90.	Akasamada Jano
		91.	Hatikhal Jano
		92.	Kantei Jano
		93.	Baharchilika Bahani, Biripadar
		94.	Baharchilika Bahani, Khirisahi
		95.	Baharchilika Bahani, Siandi
		96.	Baharchilika Bahani, Jhilimithakua
		97.	Panichua Prawn
		98.	Ginnapali Prawn
		99.	Pahari Prawn
		100.	Adanasi Mithakua Prawn
		101.	Jharakata Prawn
		102.	Abhimuhan Prawn
23.	Trinathdev P.F.C.S.	103.	Rasakudi Jano
		104.	Matia Jano
		105.	Jalimuhan Biluamari Jano

(contd....)

Sl. No	Name of the P.F.C.S.	Name of the Source Allotted
		106. Sundhimala Jano
		107. Madhara Jano
		108. Sialchatara Jano
		109. Baliachatara Jano
		110. Bagheta Jano
		111. Baligotha Jano
		112. Gobari Jano
		113. Uthapani Rasakudi Prawn
		114. Uthapani Khatiakudi Prawn
		115. Jalimuhan Biluamari Prawn
		116. Kothanikasi Prawn
		117. Heta Jano
		118. Khatiakudi Jano
24.	Taradevi P.F.C.S.	119. Dhanahara Guachopi Prawn
		120. Khanimuha Jano
		121. Sankhanadi Chingudi
		122. Chiradanasi Chingudi
		123. Khanimuhan Jano (Kurupal)
25.	Bagdevi P.F.C.S.	124. 42 Baharchilika Chingudi Khanda III, Part
		125. Bhobania Jano
26.	Narayani P.F.C.S.	126. Solakudi Jano
		127. Solakudi Chedara Jano
		128. Uparchatara Jano
		129. Talachatara Jano
		130. Barunkhati Chingudi 60%
		131. Gheemuda Chingudi 50%
		132. Solakudi Bhitar Chingudi
		133. Talabahili Talachikili Chingudi
		134. Bodiakudi Chingudi
27.	Arakhakuda P.F.C.S.	135. Manikpatana Nadi Bahani, Arakhakuda
28.	Jadupur P.F.C.S.	136. Pandanali Jano
		137. Upar Gadagauia Jano
		138. Debankhar Jano
		139. Gadagnaia Uthapani
		140. 42 Baharchilika 1/3 of 2nd part
29.	Barunei P.F.C.S.	141. Badajora Jano
		142. Sanajora Jano
		143. Baharchilika Khatisahi Gramatali
		144. Matikuda Prawn
		145. Khatisahi Prawn
		146. Panibat Prawn
		147. Kolinasi Prawn
30.	Kholamuhan P.F.C.S.	148. Galia Jano
		149. Kholamuhan Jano
		150. Tampolakona Jano

(contd....)

Sl. No	*Name of the P.F.C.S.*	*Name of the Source Allotted*
		151. Kholamuhan Galia Jano Prawn
		152. Uthapani Harinala Khanda
		153. Sialbandha Chingudi
		154. Kuanria Dian
31.	Mangala P.F.C.S.	155 Gopata Jano
		156. Kholaholi Prawn
		157. Kodiabatia Jano
32.	Uttarchilika P.F.C.S.	158 Jagadal Golara Uthapani Dian
		159. Jagadal Golara Baharchilika
		160. Baharchilika Bahani Kaudikhan
		161. Jagadal Khetandi Khati
		162. Jagadal Golara Khati
		163. Hari Das Khati
		164. Subhadrapur Nali Bohani
		165. Subhadrapur Nali (Bhaui)
33.	Chilika Jagannathpur	166. Baharchilika Bahani, Jaripada
34.	Jagulei P.F.C.S.	167. Barapania Jano
		168. Panchudera Jano
		169. Nuakhar Majhi Adanga
35.	Mamubhanja P.F.C.S.	170. Baharchilika Bahani, Dhuanl
		171. Baharchilika Bahani, Nathapur
36.	Bainsiswan P.F.C.S.	172. Baharchilika Bahani, Tentulia Pada
		173. Nathapur Kantia Jano
37.	Laxmi P.F.C.S.	174. Kampana Jano
		175. Sankha Jano
38.	Nairi P.F.C.S.	176. Baharchilika Bahani Nairi
39.	Jagamaya P.F.C.S.	177. Talagadagadia Jano
		178. Kalandilmda Uthapani
		179. Bauribada Uthapani
		180. Talagaon Uthapani
		181. 42 Baharchilika 4th Part
		182. 24 Baharchilika 1st Part
		183. Mudiratha Singitali Prawn
		184. Jamunakuda Prawn
		185. Kholanalo Uthapani
		186. Kerandia Jano
40.	Chandidevi P.F.C.S.	187. Baharchilika Bahani, Bankiapalli
41.	Man-Polangeswari P.F.C.S.	188. Baipur Nali Bahani
		189. Boat Nali Bahani
42.	Bhagabati P.F.C.S.	190. Baharchilika Bahani, Soran
43.	Badabahania P.F.C.S.	191. Balipatpur Nadi Jano
		192. Baharachilika Bahani, Bhusandupur & Balipatpur
44.	Digrisahi.	193. Baharchilika Bahani, Jaripada
45.	Gangakali P.F.C.S.	194. Baharchilika Bahani, Baulabandha
46.	Ambika P.F.C.S.	195. Knauakhat Jano.
		196. Baharchilika Bahani, Chandraput

(contd....)

Sl. No	Name of the P.F.C.S.	Name of the Source Allotted
47.	Gopinathpur P.F.C.S.	197. 42 Baharchilika 2/3rd of 2nd part
		198. Kantei Ghodasipal Jano
		199. Raipur Nali Jano
48.	Gajapatinagar P.F.C.S.	200. Pankua Jano, Bhekta Jano, Duba Jano of Laxminpur
		201. Chingudi Khanda, Gajapatinagar
		202. Binchhanapalli Prawn
49.	Kholamuhan P.F.C.S.	203. Talarihuda Jano

The basic objective of the co-operative organisation, which was set for the Chilka fishermen sometimes ago was that the primary society would receive the whole catch from individual member fishermen and finally supply the same to Apex society (CFCMS Ltd.). Till date this aim has not been achieved. The primary societies are marketing their catch at their own convenience, and as a result, of which, individual fishermen faces a great loss being exploited by the middlemen. Most of the PFCS have also not been able to give their members a better economic situation. The members still go to the private merchants and the merchants and middlemen continue to take advantage of this situation and it weakens the already weak cooperatives. As a result even though there are 68 primary societies (PFCS) in the Chilka area with 16117 members, some primaries have no activities of their own and exist only as a name.

Of the 51 active PFCS, the three biggest are—

(a) Soren Chilka varieties with 1094 members of which 1060 are active.
(b) Uttar Chilka with 1066 members of which 964 are active.
(c) Rambha with 784 members of which 729 are active.

The smallest two are—

(a) Mamu Bhanja with 11 members of which 10 are active.
(b) Baseli with 22 members of which 20 are active.

But the primary societies (PFCS) have little interest in the Apex society. Many of them exist only as a name, so that they will be able to take lease of the fishery sources. As much when a primary (PFCS) can not pay or would not like to take lease of certain fishery area, that are being leased out to another primary (PFCS) on a negotiation basis. If no other primary is interested, the Apex can decide to put it on public auction to be utilised till the end of the 3 year period. But there are exceptions like Satpada, Soren, Durga PFCS, which have a working structure and their own marketing. The members of these primaries are more active in their societies.

Primaries are sometimes unable to operate the fishery source of the lagoon on taking lease from CFCMS Ltd. due to

(a) Fisheries areas are disputed, as they are not demarcated properly in the lagoon and not surveyed by the Revenue Department. The maps that exist are not understood by all fishermen.

(b) Un-authorised fishing activities in the area by other fishermen and non-fishermen.
(c) Natural calamities and changes of environment are at times making the areas as in-operative and unproductive, even though the lease value is increased by 10% after each lease period.
(d) Conflicting interests in utilizing the fishery systems. Because of the very high price for prawns every one wants to catch more prawns. This led primaries or individuals for unauthorised conversion of a system to other than stated by the Revenue Department.

In fishery source where it is known to be conflicts, a primary (PFCS) may not want to take lease, and these conflicts are quite severe affecting many areas. Due to these conflicts, though 265 fishery sources were sub-leased in 1986-87, in 1987-88 the number of leased out fisheries decreased to only 163.

The CFCMS Ltd. had obtained 286 fishery sources from the Revenue Department for sub-leasing to the primaries as below.

Jano (Traditional enclosement fisheries)	90
Prawn (Trap fishing for prawn)	59
Bahani (Net fishing)	38
Uthapani (Fishing in shallow areas during the first rainfall in June)	6
Dian (Fishing in small shallow areas upland from some Jano areas only for non-fishermen)	93

Some keuta dominated villages have been economically strengthened even if the economic situation for most fishermen has not improved since the Apex took its form. In those villages there usually exist strong primary societies with their own marketing structures and the fishermen do not have to rely so much on the middlemen any more. Even though the income of many of the primaries have increased manifold yet, they are not paying the lease values to Govt. through the Apex with some pretext and other. They neither look to the welfare of their members nor strengthen the primaries, through which they are getting lease of fishery sources and pay the revenue to the Govt., the fact of which came to light from a recent sample survey of some PFCS as shown in Table 8.5.

The informations reported in the Table 8.5 reveal that even though some of the primaries are catching substantial quantities of fish and prawn and earning a good amount, but reluctant to pay lease value to the State Govt. Their catches are mostly sold through private traders instead of societies, though they enjoy fishing rights and lease of fisheries under co-operative fold and thereby weakening the primaries.

64 nos. of P.F.C.S. were existing upto 31.3.91 including one P.F.C.S meant for Fishermen (By profession) belongs to villages Jamuna under Bramhagiri block i.e. Jaya Maa Mangala P.F.C.S.

Table 8.5

Sl. No.	Name of P.F.C.S.	Existing member-ship	List of fishery sources taken on lease from 1988-89 to 1990-91	Qty. of fish and prawn caught (in tonnes)			Value of catch (in Rs.)	Mode of disposal of catch	Lease value paid (Rs.)
					Fish	Prawn			
1.	Keshpur	283	(1) Kachhakheda Jano	88-89	130.0	9.5	13,00,000	Sale by the members	Nil
			(2) Kalajhar Farkanasi jano	89-90	135.0	9.1	13,20,000	directly and not	
			(3) Diankudikon Balia-chatum jano	90-91	133.0	10.0	13,10,000	through primaries.	
			(4) Jallumuha Chilka Khani						
			(5) Budhibar prawn (North)						
			(6) Budhibar prawn (South)						
2.	Gangadevi	261	(1) Bahar Chilka Bahani	88-89	156.0	3.0	18,60,000	Sale by the members	2,360.00
			(2) Bathara Upper sahi	89-90	170.0	2.5	19,50,000	directly and not	
				90-91	150.0	4.0	19,00,000	through primaries.	
3.	Arkhakuda	402	(1) Manikpatna Nadi Bahani	88-89	197.2	115.5	27,18,500	Through private	
				89-90	195.5	119.2	27,65,500	merchants	12,728.00
				90-91	200.0	120.0	28,00,000	locally.	4,538.00
4.	Alandapatna	110	(1) Amrutanidhi Jano	88-89	32.1	3.1	2,70,100		
			(2) Sagakhia Kahunia Jano	89-90	34.5	3.2	3,21,500		
			(3) Kankadagada Jano, Bahani, Gramtali.	90-91	36.0	3.6	3,96,000	Through private merchants locally.	283,224.00
			(4) Chadhavanasi Chingudi Khati.						
			(5) Bali Anla chingudi khati.						
			(6) Tininalia Dian						
5.	Barunei	239	(1) Badajora jano	88-89	40.2	56.5	15,32,000		
			(2) Sanajor jano	89-90	41.5	55.2	16,03,000		

		(3) Baharchilka Khetisahi gramtali	90-91	44.4	66.6	19,98,000		
		(4) Matikuda prawn khati						
		(5) Khatisahi prawn khati						
		(6) Panibata prawn khati					50% through pvt. merchants locally. 50% through society at Balugaon.	130,000.00
		(7) Kolinasi prawn khati						
6.	Satpada	332 (1) Panchabatia Chandra-batia jano	88-89	6.75	64.2	17,50,000		
			89-90	3.25	29.0	8,26,000		
		(2) Bahar Adang jano	90-91	4.68	42.0	11,22,600		
		(3) Sednanadi Naitahal prawn khati						
		(4) Badajorda						
		(5) Serakhanda						
		(6) Barpania						
		(7) Nuakhor Majhi						
		(8) Panchudora						
7.	Mahanisa	178 (1) Ostapali	88-89	28.5	43.36	12,73,200	50% through pvt., merchants locally. 50% through society at Balugaon.	Nil
		(2) Mahnisa gramtali	89-90	24.18	32.00	10,80,200		
			90-91	30.57	38.00	16,16,032		
8.	Chaubardev	115 (1) Chandanamara Nadi Bahani	88-89	15.5	20.5	5,65,000	Through pvt. merchants locally.	13,210.00
			89-90	17.52	25.2	7,14,300		
			90-91	18.0	36.0	9,90,000		
9.	Berhampur	312 (1) Marei jano Bahani	88-89	35.2	17.5	7,72,400	50% through pvt. merchants locally. 50% through society at Balugaon.	132,228.00
		(2) Kelikuda jano	89-90	34.5	19.2	7,98,000		
		(3) Rama Bharati prawnkhati	90-91	36.0	18.0	9,60,000		
		(4) Chaliskanda prawn khati						
		(5) Sukurapa prawn khati						

(contd.....)

Sl. No.	Name of P.F.C.S.	Existing membership	List of fishery sources taken on lease from 1988-89 to 1990-91	Qty. of fish and prawn caught (in tonnes)			Value of catch (in Rs.)	Mode of disposal of catch	Lease value paid (Rs.)
					Fish	Prawn			
10.	Jadupur	79	(1) Pandanali jano	88-89	14.2	13.7	3,87,600	50% through pvt.	9,003.00
			(2) Upargadagadia	89-90	15.5	12.5	4,05,000	merchants locally.	
			(3) Debankhor jano	90-91	12.0	14.4	4,80,000	50% through society	
			(4) Gadagadia uthapani					at Balugaon.	
			(5) 42 no. Bahar Chilka prawn						
11.	Kanakdurga	279	(1) Ginapali jano	88-89	26.0	45.4	11,68,000	Through private	99,797.00
			(2) Garagasi Jano	89-90	28.5	48.2	13,06,000	merchants.	
			(3) Deuli Jano	90-91	30.0	50.0	14,50,000		
			(4) Khirisahi Jano						
			(5) Akasmuda jano						
			(6) Hatikhol jano						
			(7) Kantei jano						
			(8) Bahar chilka Bahani						
			(9) Khirisahi Bahani						
			(10) Biripadar Bahani						
			(11) Jhilikua prawn						
			(12) Panichua prawn						
			(13) Ginapali prawn						
			(14) Pohori prawn						
			(15) Jharkata prawn						
			(16) Adanasi Mithkua prawn						
			(17) Abhimuha prawn						

Table 8.6

S.No.	*Name of the P.F.C.S.*	*Date of Registration*	*Remarks*
	List of P.F.C.S. availed lease upto 1990-91		
1.	Rambha	31.7.59	
2.	Chamunda	31.7.59	
3.	Gajapatinagar	31.7.59	
4.	Keshpur	31.7.59	
5.	Jogamaya	18.8.59	
6.	Narayani	18.8.59	
7.	Ambika	18.8.59	
8.	Jagulai	18.8.59	
9.	Durga	18.8.59	
10.	Kalupada	18.8.59	
11.	Gopinathpur	25.8.59	
12.	Jageswari	25.8.59	
13.	Trinathdev	25.8.59	
14.	Laxmi	25.8.59	
15.	Kanakadurga	25.8.59	Liquidated during 1/94.
16.	Berhampur	7.12.59	
17.	Mangala	18.4.60	
18.	Satapada	6.1.61	
19.	Harachandi	6.11.61	
20.	Barunei	10.12.62	
21.	Arakhakuda	18.3.63	
22.	Sorana Chilka	17.6.63	
23.	Baradi	17.6.63	
24.	Bhusandapur	12.6.63	
25.	Uttar Chilka	12.6.63	Liquidated during 2/94.
26.	Nairi	17.6.63	
27.	Chandidevi	12.6.63	
28.	Bhagabati	24.6.63	
29.	Bainiswan	28.8.63	Liquidated prior 1992.
30.	Pathara	28.8.63	
31.	Gangakali	5.8.64	Liquidated prior 1992.
32.	Mahamaya	25.6.71	
33.	Chilka Jaganathpur	3.8.71	
34.	Taradevi	11.11.71	
35.	Jadupur	19.6.74	
36.	Jenamani	30.1.76	
37.	Bagdevi	9.2.76	
38.	Badabahaniasahi	4.1.78	Liquidated during 1/94
39.	Padmadevi	26.7.78	
40.	Mahinsa	19.12.81	
41.	Alandapatna	1.1.82	

(contd....)

S.No.	Name of the P.F.C.S.	Date of Registration	Remarks
42.	Sabulia	26.11.82	
43.	Gangadevi	25.5.51	
44.	Chilka Digirisahi	11.1.83	
45.	Bhudhima	12.10.83	
46.	Maa Polangeswari	22.12.83	
47.	Chourabardev	30.4.85	
48.	Mamubhanja	10.10.85	Liquidated prior 1992.
49.	Kholamunnha	4.12.85	
	List of P.F.C.S. Non-availed lease		
50.	Haimabati	4.1.60	
51.	Chilka	22.3.61	Liquidated during 1/94.
52.	Uttarayani (Gabapadar)	7.6.62	
53.	Mangalajodi	17.6.63	Liquidated during 1/94.
54.	Chandi	17.6.63	
55.	Bhagabata	28.8.63	
56.	Kubudi	17.1.65	
57.	Mahabir	28.8.71	
58.	Refusee	29.1.79	
59.	Ugratara	13.2.79	
60.	Maa Bhargabi	2.4.81	
61.	Basudev	24.1.84	
62.	Jai Maa Mangala	8.2.85	Non Fisherman.
63.	Uttarayani (Hatabaradi)	18.5.87	
64.	Parbatipur	17.11.90	

Out of above 64 nos. of P.F.C.S. only 49 nos. of P.F.C.S. have availed lease of 205 fishery sources of Chilka lagoon through Central Fishermen Cooperative Marketing Society Ltd., Balugaon, during 1988-89 to 1990-91 stated earlier.

After 31.3.91 till 31.3.94 following 11 nos. of new P.F.C.S. have been registered in Chilka area.

(1)	Baseli P.F.C.S. at Patuabahaniasahi of Balipatpur under Tangi Police station in Khurda Dist.	This P.F.C.S. has been registered vide Regd. No 71/28.4.91 for fishermen by caste. This society was availing Bahar Chilika bahani and Balipatur River Sairat sharing the sub-lease allotted in favour of Badababahaniasahi P.F.C.S. which has been liquidated.
(2)	Maa-Kalijai Fisherwomen Dry Fish Production & Marketing C.S. At Bhusandapur in Khurda District.	This Society has been registered with registration No.72 dt.29.4.93 for fisher-women with objects of production and Marketing of Dry Fish and self employment.

(3)	Meghambari P.F.C.S. At Bankiapalli under Balugaon P.S. in Khurda Dist.	This society has been registered with Regd. No 73 dt.3.7.91 for fishermen by caste. This society has been comprised from Chandidevi P.F.C.S. and is availing Chilka lease of Bahar Chilka Bahani sharing the allotment of sources made in favour of Chandidevi P.F.C.S.
(4)	Maa Majhiani P.F.C.S. At-Mangalajodi with area of operation confinding to Mangalajodi, Bhusandapur, Balinasi & Chandpur in Khurda district.	This society has been registered with registration No.75 dt. 10.3.92 for fishermen the members of the society are catching from Bahar Chilka Bahani without allotment of any fishery source.
(5)	Maa Mohalaxmi P.F.C.S. At-Baulabandha under Banpur P.S. in Khurda Dist.	This society has been registered with registration No76. dt. 18.3.92 for fishermen. The society has been comprised from Laxmi P.F.C.S. at Nairi whose area of operation is the same. The members of the said P.F.C.S. are availing Chilka lease and Bahar Chilka Bahani through Laxmi P.F.C.S.
(6)	Kalijai P.F.C.S. At-Balugaon in Khurda Dist.	This society has been registered with registration No.2 CTC(FY) 11.11.92 for the fishermen of Kumandal village. This society has been comprised from the area of operation of Taradevi P.F.C.S. The members of the society are availing Chilka fishery source through the sub-lease of area allotted in favour of Taradevi P.F.C.S.
(7)	Ma Taratarini P.F.C.S. At-Gorapur samil Balabhadrapur under Brahmagiri P.S.	This society has been registered with registration No.5 CTC (Fy) dt.29.1.94 for fishermen by profession. The members of the society are availing Chilka fishery source on deposit of lease value with Tahasildar, Puri since 7.2.75.
(8)	Maa Baliharachandidevi P.F.C.S, At-Anua under Krushna Prasad P.S.	This society has been registered with registration No.74 dt. 7.9.91 for fishermen by profession of Anua village. The society has been registered basing upon the allotment of Chilka fishery source of Ac.102.57 during the year 1988-89 to 1990-91 made in favour of the society by the Tahasildar Krushnaprasad.
(9)	Maa Mangala P.F.C.S. At-Tichhini under Krushnaprasad P.S.	This society has been registered with registration No.3 CTC(FY) 11.11.92 for the fishermen by profession of the village Tichhini. Registration has been made in consideration of allotment of

	Chilka fishery source of Ac. 244.41 made in favour of the society by the Tahasildar Krushnaprasad on deposit of Rs. 1,95,052 made towards first instalment of lease value.
(10) Maa Padukai P.F.C.S. At-Alanda.	This society has been registered with registration No.4 CTC(FY) 10.1.94 for the fishermen by profession of Alanda village. Registration has been made in consideration of allotment of Ac.44.93 in favour of village head Sri Bainsidhar Pallai and the member of society by the Tahasildar Krushnaprasad on 28.1.92 on a deposit of Rs. 22,000 towards instalments of lease value.
(11) Maa Mangala P.F.C.S. At-Patharganja.	This society has been registered with registration No.6 CTC(FY) dt. 31.1.94 for the fishermen by profession of the village Patharganja. Registration has been made in consideration to the allotment of Chilka fishery source of Ac. 196.42 made in favour of the society by the Tahasildar Krushnaprasad on 28.1.92.

Out of these 11 nos. of P.F.C.S, one P.F.C.S is organised by Fisheries Women to take up dry fish trade and 4 nos. P.F.C.S. meant for fishermen (By profession) and the rest 6 nos. of P.F.C.S. are organised by the fishermen by caste on capture fisheries (Chilka Bahani) and thus the total nos. of P.F.C.S. come to 75 till 31.3.94.

8 P.F.C.S. have been put under liquidation on the ground of violating the Rules and provision of O.C.S Act. and Rules; specifically relating to violation of Cooperative Election Rules 1992.

1. Ganga Kali	The society was under liquidation due to defunct condition prior to inforcement of amendment of O.C.S. Act, 1991.
2. Mamubhanja	The society was under liquidation due to defunct condition prior to inforcement of amendment of O.C.S. Act, 1991.
3. Bainsisona	The society was under liquidation due to defunct condition prior to inforcement of amendment of O.C.S. Act, 1991.
4. Kanaka Durga	The society violated the provisions of O.C.S. Act and election rules for constitution of Board of Management due to non-availability of records.
5. Chilka	The society violated the provisions of O.C.S. Act and election rules for constitution of Board of Management due to non-availability of records.

6.	Mangalajodi	The society violated the provisions of O.C.S Act and elections rules for constitution of Board of Management due to non-availability of records.
7.	Badabahaniasahi	The society violated the provisions of O.C.S. Act and election rules for constitution of Board of Management due to non-availability of records.
8.	Uttara Chilka	The society violated the provisions of O.C.S. Act and election rules for constitution of Board of Management due to non-availability of records.

Fishing with net in deeper Chilka

Ponds constructed in marginal area of the lagoon for prawn culture

Enclosing Jano area with pen materials for prawn culture

Operation of a battery of traps for catching prawn with in the ghery bundh

Pen culture enclosure in Chilka lagoon

Fixing poles and pen materials for prawn culture within the lagoon

Obstruction of navigational passage by the wooden poles and net for prawn farming

Obstruction of Bhobania nala by pen materials for prawn farming

9

Fish Trade in the Lagoon

The development of fresh fish trade of the lagoon has been initiated due to the enterprise of outside merchant from Bengal, who adopted the system of taking the lease of lagoon fisheries from the owners and sub-leasing the same to the local fishermen at higher rates reserving monopoly rights, receiving the entire catch from their respective fisheries since 1920. The export figure from the lagoon, before second world war was on an average 50,000 maunds annually, most of it having been sent to Calcutta.

With the establishment of an ice factory at Kaluparaghat, the export of fresh fish from the lagoon developed rapidly making the Kaluparaghat the largest export centre of the lagoon.

In the beginning the merchants utilised the services of some fishermen to collect catches from the fishing grounds and bring them to the assembly points near the rail heads. These fishermen in course of time gave up fishing and took the profession of commission agents. They weigh and took fish from the fishermen in the fishing grounds keeping sufficient margin for any shortages due to future loss in weight and weigh out the same to the merchants depots at the assembly points.

As many as 300 heavy country boats, using bamboo mats as sails, were engaged to transport fish some time covers a distance of 32 kms. or more. If the wind was favourable the fish could be booked to Calcutta on the same day, or else it has to be left over for the night, for which fishermen were paid lesser price for fish brought late.

Usually fish were packed in horizontal layers with fish and ice alternately in the ratio of 1:1 as fish and ice in summer and 3:2 in winter, leading more than 40 per cent spoilage on arrival in Calcutta. The fish was auctioned publicly at Howrah and the *adatdars* (the principal merchants) used to charge 3.12 per cent if no money was advanced, double otherwise from the auction value of fish sent.

Out of Rs.11 lakhs as pre-war gross income of the lagoon from fisheries, the fishermen for all their troubles used to receive between Rs.1 to 2 lakhs per year. During the war the income from the lagoon fisheries went up to nearly Rs.20 lakhs per annum. By introducing a system of control of prices both at the source as well as at the retail market for a portion of the supplies it was possible to apportion 40 per cent of this amount to fishermen.

This fact, that the fishermen to get due reward for his labour, the Central Fishermen Co-operative Marketing Society Ltd. was formed at Balugaon, in one of the fish assembly point of the lagoon to successfully market its catches in Calcutta.

But individual merchants operating in the lagoon always maintain their upper hand than that of co-operative society and the percentage of fish exported by the merchants were also higher than that of the co-operative societies. Between 266 to

281 individual merchants and 166 to 231 cooperative societies were operating between 1949 to 1954 handling 55.3 to 64.4 and 35.6 to 44.7 percentage of fish export trade of the lagoon during the period. Fresh fish of the lagoon was then exported only from rail heads like Kaluparaghat, Kuhuri, Gangadharpur, Balugaon, Khallikote and Rambha of which Kaluparaghat and Balugaon were the main centres handling 60 to 65 per cent of the fish traffic.

Inspite of the creation of the co-operative system in 1959 (CFCMS Ltd.) the main part of the fish is traded through the private market. It seems to be difficult to change this as the private market is the traditional way for selling fish. The private ownership over parts of the lagoon and the merchants and traders monopoly over the fish market, which grow to prominence during the British rule, created this dependence. Though in bye-laws of the CFCMS it was stated that all the fish caught should be sold through the Apex, yet this is not done in most cases, not even through primaries, and there is no provision in the bye-laws to punish the offenders. Some of the prosperous societies though, impose a fine if their members do not sell a stated part of their catch through the primary's marketing channel. The percentage of the fish sold through the co-operative system has varied over the years and has only been about 10 per cent on an average. During 1968-69, fishes and prawn from the lagoon were exported by the Co-operative societies and individual merchants leaving nearly 30 per cent for local consumption inside the State. A total of 2423.95 tonnes (68.51% of the total fish catch) from the lagoon was exported to Calcutta during 1968-69.

Marketing of lagoon fish through co-operative societies further dropped out 4.74 per cent (322 tonnes) of total catch (6862 tonnes) in 1987-88 through 16 Co-op. societies out of 68 primaries as shown in Table 9.1.

Table 9.1

PFCS	*Qty. of fish marketed (in tonnes)*
Satpada	45.0
Mahisa	40.0
Durga	31.0
Soran Chilka	70.0
Bhusandpur	3.0
Padmadevi	45.0
Badabahani Sahi	1.5
Narayani	27.0
Keshpur	1.0
Kolamohan	6.2
Mahamaya	1.2
Pathara	40.0
Gangadevi	8.0
Trinathdevi	8.0
Ma-jaimangala	1.5
Total	321.8

The export of prawn from the lagoon outside the country was first started in 1970-71 and with the increasing export potentialities of the lagoon prawn, eight prawn processing plants were established along the periphery of the lagoon for processing and export of prawns. In 1990-91, the total fish and prawn catch of 4273.3 tonnes from the lagoon were marketed in the following manner.

(i)	Export of fish to outside state	2134 tonnes
(ii)	Export of prawn to foreign market	180 tonnes
(iii)	Consumption of fish & prawn inside the state	1745 tonnes
(iv)	Consumption of fish and prawn by local inhabitants in and around the lagoon	214 tonnes.

Price of Chilka Fish and Prawn

The producer (fishermen) of the lagoon used to get only 10 to 40 per cent of sale price of prawn and fish. The bulk share was going to the wholesaler (adatdar) commision agents.

The average pre-war price of Chilka fish and prawn per maund (37.5 Kg.) were:

	Price in Rs. per maund (37.5 Kg.)		
Mugil cephalus	5.50	-	6.50
Mugil subviridis	4.00	-	6.00
Polynemus indicus	4.00	-	6.00
Hilsa ilisha	5.50	-	7.50
Lates calcarifer			
(a) Big sized	16.00	-	20.00
(b) Small sized	12.50	-	15.00
Chryophrys datnia	3.00	-	4.00
Pseudosciana coiber			
(a) Big sized	6.50	-	8.00
(b) Small sized	3.00	-	3.50
Prawn (Big sized)	7.50	-	10.00
(Small sized)	3.50	-	5.00

The fishermen of the lagoon used to get three fold higher price than that of pre-war period during 1944-46.

	Price in m. per maund (37.5Kgs.)		
Mugil cephalus	13.00	-	15.00
Mugil subviridis	12.00	-	12.50
Polynemus indicus	15.00	-	30.00
Hilsa ilisha	13.00	-	15.00
Lates calcarifer			
Big sized	40.00	-	60.00

Small sized	12.50	-	15.00
Chryophrys datnis	12.00	-	16.00
Pseudosciana coiber			
(a) Big sized	25.00	-	32.50
(b) Small sized	12.50	-	15.00
Engraulis purava	10.00	-	15.00
Prawn (a) Big sized	25.00	-	30.00
(b) Small sized	7.00	-	10.00

The merchants used to incur Rs. 22.12 for packing 5 maunds of fish with ice including loading them into railway wagon during 1940 on the following items.

Six blocks of ice	13.50	For packing
Carriage of ice	0.50	
Cost of 5 baskets at one-third rate as each basket being used thrice	0.75	
Three blocks of ice	6.75	For supplying to collecting agents.
Carriage of ice	0.25	
Carriage, weighment & loading of fish	0.50	

The price of Chilka fish gone up gradually due to increased export outside the state.

With the export of prawn to the foreign market and increasing demand of Chilka fish in domestic market outside the state, the price structure exhibited sudden increase in all varieties of prawn and fish of the lagoon and the present average price (1991) is shown in Table 9.2.

Table 9.2

Varieties	*Price in Rs. per Kg.*
Mullet	20.00
Penaeid prawn	100.00 (40 count)
Medium sized prawn	40.00 (100 count)
Small sized prawn	20.00 (160 count)
Hilsa	20.00
Clupeids	12.00
Perches	15.00
Threadfins	12.00
Catfish	10.00
Beloneforms	8.00
Etroplus	10.00
Misc.	5.00

10

Aquafarming in the Lagoon

Culturing prawns in the lagoon is relatively new activity to the area. Due to the big demand and high price of prawns, aqua-culture of prawn has become an issue in the lagoon. Prawn culture could make it possible to divert a lot of fishermen from the capture fisheries in the lagoon.

Rain-fed pond culture of prawns was introduced in the area in 1983 and in 1991 prawn farms occupied over 4000 ha. of shoreline and shallow areas of the lagoon. These rain-fed ponds of 0.11 to 0.53 ha. having 1-2m. depth of water, retain 0.8 to 1.0 m. of water during culture period and dry out during April, June, giving a yield of *P. monodon*, 156.2 to 608.4 Kg./ha./crop. Though this would lead to socio-economic improvements of the local people; yet prawn farming can cause significant impacts on the environment, wherever it is carried out and in a confined environment, such as, in Chilka lagoon; and these impacts could be very serious.

Prawn culture, due to its strong, self-sustaining and high economic value, which earns better than average wages for the local people is expanding very rapidly, creating conflicts with traditional resource users and significant water and land use changes around some parts of the lagoon. As a result, the prawn farming activities are taking place in Jano-areas. These activities are not legally granted. Because of the very high price of prawns, the villagers and individuals make their own decisions about in what way they use the jano-area. This is also true for some fishermen primary societies, even if they are affiliated to the Apex society. This activity is creating both human and ecological conflicts in the lagoon.

A part of Kholamuha jano (435 acres) comprising 300 acres was stocked with 12 lakhs of prawn seed valued to Rs.120,000 in first week of July, 1991, by erecting pen of bamboo thatta over low mud wall. After 60 days of stocking, from 23.8.91, 7 tonnes of 20, 30, 40 count prawns were harvested valued to 7 lakh of rupees. Another 4.6 tonnes of prawn was anticipated from the area till final harvesting giving a total income of Rs.11.6 lakhs against their expenditure of Rs.3.5 lakhs giving a net profit of Rs.8.10 lakhs from the area. The production thus obtained from the area is calculated to be 96.6 kgs./ha.

Prospects of Prawn Culture in Chilka Lagoon

The exposed area of 259 sq. km. between high and low water level can be properly utilised for aquaculture purposes mainly for tank and pen culture of brackishwater prawns. Of the total 25900 ha. (259 sq. km.) exposed area, 5005 ha. has been demarcated for the development of prawn farms under various centre-State Schemes (Status Report, BFDA, 1985). This forms 19% of the total available area and an another 2000 ha. can be utilised for pen culture purposes which together

Table 10.1 : Prawn Farming in Chilka Lagoon

Name of district	*Name of block*	*Area identified in ha*	*Area under prawn culture (Ha)*		*Production of prawn from culture per ha.*							
					Ganjam district				*Puri district*			
					1989-90		*1990-91*		*1989-90*		*1990-91*	
			In pond	*In ghery*	*Pond*	*Ghery*	*Pond*	*Ghery*	*Pond*	*Ghery*	*Pond*	*Ghery*
Puri	Brahmagiri	2232.33	340.22	358.00	637 Kg. in 2 crops	200 Kg. per crop	150 Kg. in 2 crops	54 Kg. per crop	738 Kg. in 2 crops	200Kg. per crop	590 Kg. in 2 crops	210 Kg. per crop
	Puri Sadar	1.10	—	—								
	Krushna Prasad	3998.10	416.91	1813.86								
	Chilka	1238.38	34.40	—								
	Tangi	897.50	131.62	—								
Ganjam	Keshpur	115.78	54.66	26.04								
	Pathara	24.42	7.92	3.20								
	Langleswar	41.18	10.89	8.08								
	Sabulia	2.16	—	—								
	Rambha	2.00	—	—								
	Kanaka	17.68	—	14.00								
	Ramgarh	743.62	202.32	162.65								
Total		9314.25	1198.94	2345.83								

Ghery Bundh in Marginal Area of the Lagoon Constructed for Prawn Farming

forms 27% of total area. These areas have to be earmarked based on the retention of 2' to 4' water level atleast for 5 to 6 months period either for tank or pen culture.

The water in Chilka has uniform salinity perferably within 10-20 ppt. with little seasonal variation. The lake is shallow and eutrophic in nature, which offers better conditions for farming purposes. The soil is of mostly sandy-clay in nature which forms ideal subsratum for prawns to thrive.

Pond Culture

As such there is no tidal influx in Chilka Lake and so confined pond management is practised. The ponds got filled with percolation water, when the water level in Chilka Lake raises during rainy season. Likewise, the water level in pond gradually decreases with the gradual decrease of Chilka water towards summer. So, culture practice usually coincides with the raise in water level in pond depending on the intensity of rains. The great advantage of pond culture in Chilka is, the tank goes dry during summer, thus improving the fertility status of soil. A few tanks which retains little water even during summer poses problem of deterioration in water and soil quality.

The advantages and disadvantages of confined pond culture are as follows:

Advantages

(a) Predator entry is controlled fully.

(b) Changes in physico-chemical parameters of water occurs gradually as there is no water entry from outside except on rainy days.

(c) The tank goes dry during summer exposing the soil which increases natural fertility.

Disadvantages

(a) Application of manures as well as feeding are to be done in low quantity, otherwise the water quality will be deteriorated.
(b) Stocking of prawn seeds should be at minimum which in turn reflects on low yield.
(c) Similar farming methods are to be continued without any scope for further improvement.
(d) Harvesting of prawns will be at low quantity every day which continues atleast for a month.

Pond Management and Production

The ponds retain percolation water, when the water level in Chilka rises and thus coincides with the rainy season. The pond was treated with lime (100 kg./acre) before water rises in the tank. Application of organic and inorganic manures followed after a few days from the day of application of lime. Production of phyto and zoo-plankton reaches a peak after a lapse of one week, during which the seeds collected from natural sources are stocked. The stocking density varies in relation to size of seeds, but comparatively less than that stocked in tide-fed farms. The feed provided during culture is either crushed meat of snails or artificial pelletised one (manufactured by Tata Oil Mills, Madras) at the rate of 5-10% of biomas of the prawns.

Fringe Area of Chilka Lagoon Converted into Prawn Culture Ponds

The water level in the pond starts decreasing gradually after completion of rainy season, particularly from October onwards. Selective harvest of only bigger prawns starts from October coinciding with the decrease in water level by means of bamboo traps. Harvesting is continuous process which helps faster growth of remaining stock by thinning out the population.

A few farmers also practise continuous stocking and continuous harvesting, depending upon the water level in the pond and the existing prawn stock. This method works better as the crop-wise pattern can not be practised due to fluctuation of water level, in pond depending upon Chilka water level which in turn depends upon intensity of rain. The production of prawns in confined pond culture methods usually varies from 60 kgs. to 150 kgs. per acre during the water retention period.

Prawn Culture in Jano-grounds

Prawn farming has now been taken up in both janos and other more or less enclosed water areas. This is a recent development and it is done by private entrepreneurs. Farming in this way is not legally sanctioned. The janos used for this enterprise are shallow and situated close to land.

The areas were in some cases janos leased out through the Apex and others were earlier janos, but have now become defunct. A big area, 300 ha. of water had been enclosed when a road was constructed straight through a productive jano. The area has some conflicts and no jano-activity is there today. The resulting confined water-area was now used for extensive prawn farming. A big area like this was not stocked at a very high density, 5000 seed/ha. The income from the year of stocking was up to November Rs.500000. The area had needed hardly any investments apart from the seed. It was totally extensive, i.e. no feeding and no fertilizer were being used.

Most other jano-areas used for farming were not that big. Two areas visited were 4.9 ha. and 12.2 ha respectively. These were still leased out as jano systems by the Apex, but the primaries did not use the areas themselves. The primaries had not enough money for investments in prawn-farming. They subleased it to interested private investors at a higher lease-value. The lease for the smaller area was Rs.5000 for one year and for the big Rs.15,000 for a 3 year period. The smaller area was inexpensive because it was very well protected against bad weather conditions. For the bigger area it had been necessary to invest in an embankment costing Rs.30,000. This area was said to be very productive, so after the end of the 3 year lease they expected the lease-value to increase sharply, perhaps even upto Rs.100,000 year.

The farming strategy seemed to be based on trial and error to a great extent as they adopted the same technique as is used, when culturing prawns in rain-fed ponds, even if the ecology is not at all the same for these two culture systems. The 4.9 ha. jano was closed by a mud wall with only a smaller opening left open to the lake. In this opening they put a bamboo-screen and some nylon-netting of fine meshsize, i.e. mosquito net. This is all done before the monsoon, so when the water comes, no unwanted species can enter the farm, as they would be either competitors or predators. In 1988, they had not been able to close this jano until August because of some conflicts. In the area many different species occurred. The species causing most damage to the prawns were *Lates calcarifer, Eleutheronema tetradactylum,*

Therapon jarbua and *Plotosus canius*. Birds are also consuming a lot of prawns, mostly it is different species of egretts and cormorants.

The water outlet could be closed totally, not to let too much water out during very low water level in the lagoon. The 4.9 ha. area was stocked with 30,000 *P. monodon* and 3 000 *P. indicus*. The stocking did not take place immediately after the area was filled with water as it then had too low salinity, so they had to wait some weeks. Most of the farmers of these illegal farms bought seed collected locally in Chilka, (*P.indicus* only from Chilka). Some farmers also bought from the collection center in Paradeep. There the seeds were graded and only *P. monodon* were collected. The catching of the prawns at harvest is done traditionally by the use of traps and leaderlines. The traps are set in the evening and emptied early in the morning. This was going on until the area was becoming dry in December or January. Not very much time is needed to empty the traps. Most time is spent on guarding the area, 24 hours a day. In the 4.9 ha. area 3 guards took turns on guarding and for the 12.2 ha area 6 guards were employed. The feeding and fertilizing also takes some time. Feeding was done without knowing so much about it. It was a mixed and not very regular diet, consisting of snailmeat, henfeed, ground nut oil cake and fish-meat. Most of the food was bought locally, when needed. The use of fertilizer should be avoided as there will be a more or less continuous water exchange with the water outside the farm. Much of the nutrients will just leak out to the lagoon and might cause ecological disturbances and will not increase the productivity of the jano very much. Fertilizer was bought from Puri once to twice a month.

This illegal prawn farming in Chilka lagoon expanded very rapidly in subsequent years by large-scale subletting of the leased out fishery sources by the primary fishermen cooperative societies. Massive illegal encroachment of fishery sources by the non-fishermen and outsiders, as well as widespread conversion of traditional fishery sources like Jano, Dian, Uthapani, Bahani and the navigational channel into prawn culture ponds or net-enclosure gheries took place as shown in Table 10.2.

The use of the jano-areas for prawn farming might be a good idea from economical point of view; but it is essential that it is done in a way not destroying the Chilka ecology. The activity needs to be introduced and supervised by people with ecological knowledge. The relatively low salinity in the jano is too low for profitable prawn production. It must, of course, also be legally sanctioned.

Because of the shallowness of the lagoon and the economic pressures in the area for expansion of prawn farming industry, controlling and directing expansion of this industry in the lagoon has become a high priority.

Although prawn farming can cause significant environmental impacts, there are many areas where this activity takes place with minimal or acceptable impacts, depending on the technology that is used and the development plan that has been followed. Most environmental impacts caused by prawn farming also result from a specific type of technology used—intensive and semi-intensive prawn farming. Impacts resulting from this type have often through speculation been incorrectly associated with less intensive types of prawn farming.

Table 10.2

Sl. No.	Name and location	Location of culture area (ha.)	Total investment (Rs.)	Year of operation	Qty. of prawn seeds stocked	Harvesting		Salinity ppt.
						Quantity (Kg.)	Income (Rs.)	
1	2	3	4	5	6	7	8	9
Krushnaprasad Block								
1.	Gopinathpur PFCS, Raypur.	Raypurnali (64.00)	80000	90-91	60000	Flood affected		5 to 12
2.	Jadupur PFCS	Panda jano (80.00)	—	90-91	—	Flood affected		5 to 12
3:	Villagers, Siara. Siara.	Sairaganta (20.00)	—	90-91	—	Flood affected		5 to 12
4.	Villagers, Panaspada.	Bhoisahi Khama ((20.00)	—	90-91	—	Flood affected		7 to 15
5.	Villagers, Panaspada.	Datai Chatara (16.00)	270000	90-91	500000	3575	442750	5 to 12
6.	Villagers, Panaspada.	Gangapania (12.00)	—	90-91	—	—	—	5 to 12
7.	Villagers, Panaspada.	Lenkasahi Dian (4.80)	55000	90-91	116400	480	54100	5 to 12
8.	Villagers, Panaspada.	Suanpadar Jano (20.00)	136000	90-91	280000	2420	269700	5 to 15
9.	Villagers, Panaspada.	40.00	170000	90-91	150000	300 Flood affected	45000	3 to 15
10.	Villagers, Panaspada.	8.00	40000	90-91	—	Flood affected		3 to 12
11.	Gangadharpur PFCS, Dahikhia.	Gangadharpur Nali (20.00)	120000	90-91	—	Flood affected		3 to 12

(contd....)

1	2	3	4	5	6	7	8	9
12.	Gangadharpur PFCS, Dahikhia.	Nadiakhia Jano (28.00)	180000	90-91	200000	2400	264000	3 to 12
13.	Villagers, Nuagaon	Nuashere (10.00)	120000	90-91	272597	561 Flood affected	65810	5 to 15
14.	Rama Ch. Pradhan & others, Nuagaon.	Private land (7.20)	82000	90-91	60000	Flood affected		2 to 12
15.	Panchu Mohapatia Nuagaon.	Private land (2.40)	—	90-91	—	Flood affected		5 to 12
16.	Ram Ch. Pradhan, Nuagan.	Private land (2.40)	50000	90-91	50000	435 Flood affected	52000	5 to 12
17.	Indramani Jagadev, Gambhari.	Private land (14.00)	200000	90-91	250000	300 Flood affected	33000	3 to 15
18.	Rama Ch. Mansingh, Gambhari.	Private land (14.00)	190000	90-91	230000	515 Flood affected	57800	3 to 15
19.	Madhusudan Mohanty, & others, Gambhari.	Private land (4.80)	68000	90-91	110000	208 Flood affected	19160	3 to 15
20.	Satyapriya Mohanty & others, Gambhari.	Private land (4.80)	75000	90-91	170000	670 Flood affected	63950	3 to 15
21.	Narayan Mohanty & others, Gambhari.	Private land (10.00)	—	—	—	Flood affected		3 to 15
22.	Villagers, Gambhari, Gopalpur.	Ghata Gheri (4.80)	125000	90-91	150000	2235	247400	2 to 12
			1961000		3238997	14099	1614870	

N.B. Except above gheries (22), other gheries were affected by the flood for which culture has not been done.

1	2	3	4	5	6	7	8	9
Sadar Block								
1.	Pokakhia, Siandi	249.21		1991	2000000	4196	4,27,000	
2.	Jankhikudi, Talatola	52.00		1991	500000	1366	1,36,600	
3.	Arakhara, Talatola	60.00		1991	500000	2433	2,43,300	
4.	Patanasi Jano, Patanasi	20.00		1991	200000	2396	2,46,200	
5.	Priluamani, Naugaon	20.00 (Pen culture)		1991	80000	1708	1,70,800	
6.	OMCAD, Ramalenka	130.00		1991	390000	2193	2,16,000	
7.	OMCAD, Nandala	129.00		1991	360000	1000	1,00,000	
8.	OMCAD, Paikarapur	130.00 (Pen culture)		1991	360000	900	90,000	
9.	Chatana, Udayagiri	40.00		1991	800000	320	32,000	
10.	Bramhagikhera, Udayagiri.	15.00		1991	400000	500	50,000	
11.	Bidoikhera, Budhibara(Pvt)	32.00		1991	100000	850	85,000	
12.	Baligotha, Kandeswar	60.00		1991	250000	1250	1,25,000	
13.	Kandeswar Jani, Kandeswar	30.00 (Pen culture)		1919	100000	650	65,000	
14.	Baudaban Sagar, Malud	10.00		1991	50000	250	25,000	
15.	Kadua Jani, Ramalenka	20.00 (Pen culture)		1991	90000	500	50,000	
16.	Arakhara Nuagaone	54.00		1991	150000	1816	1,81,600	
17.	Kadua Nandila	20.00 (Pen culture)		1991	80000	750	75,000	
	Total				64,10,000			

(contd....)

1	2	3	4	5	6	7	8	9
Khallikote Block								
1.	Bharmara Dalai, Near Railway Road No.336/6 to 336/10 Chilka.	2.40	7500	90-91	28800	336	40000	12 to 22
2.	Dharani Prushty, Patra Unsurveyed Chilka-1.	3.20	9900	90-91	38400	464	49000	12 to 22
3.	Trinath Jally Patra, Kanika site	8.00	24000	90-91	96000	1120	112000	12 to 22
4.	Khetramohan Bela Patra, Unsurveyed Chilka-1.	6.00	18000	90-91	72000	840	90000	12 to 22
Ganjam Block								
5.	K.K. Patra, Site at Rayabandh.	6.00	18000	90-91	72000	870	88000	12 to 22
6.	Gopinath Naik, Site at Prayagi.	3.00	9000	90-91	36000	435	45000	12 to 22
7.	Balaram Naik, Site at Prayagi.	68.00	204000	90-91	816000	9520	102000	12 to 22
Khallikote Block								
1.	Kaibalya Behera Belapada Site at Gopakuda, Chilka-1 Unsurveyed area.	2.80	7500	91-92	33600	398	40000	12 to 22
2.	Rama Ch. Prusty, Patna, Chilka-1.	5.00	15000	91-92	60000	700	70000	12 to 22
3.	Dandapani Behera, Langaleswar, Chilka-1	3.90	10500	91-92	46800	546	60000	12 to 22

(contd....)

1	2	3	4	5	6	7	8	9
4.	Kumuda Behera, Patna, Chilka-1	10.00	30000	91-92	120000	1400	140000	12 to 22
5.	Sasibhusan Behera, Sabulia, Chilka-1.	2.00	6000	91-92	24000	280	30000	12 to 22
6.	Hari Jally, Sambal, Chilka-1	7.00	21000	91-92	84000	980	98000	12 to 22
7.	Subarna Behera, Gourangapatna, Chilka	4.00	12000	91-92	48000	560	60000	12 to 22
8.	Ladu Jali, Gokharkuda	7.00	21000	91-92	84000	982	102000	12 to 22
Ganjam Block								
1.	Sankarsan Dalai, Prayagi.	73.00	173000	91-92	876000	10220	1022000	12 to 22
2.	Ranka Dalai, Prayagi	7.00	21000	91-92	84000	982	98200	12 to 22
3.	Khaga Das, Gajapatinagar, M-Laxmanpur, Bhimpur.	16.00	48000	91-92	192000	2240	224000	12 to 22
4.	Natabar Behera, Gajapatinagar, M.Laxmanpur, Bhimpur.	4.80	13500	91-92	57600	672	68000	12 to 22
5.	Mohan Malik, Gajapatinagar, Laxmanpur, Bhimpur.	1.28	4000	91-92	15500	180	20000	12 to 22
6.	Adi Das, Gajapatinagar. M-Laxmanpur, Bhimpur.	1.20	4000	91-92	15500	150	15000	12 to 22
7.	Sarat Ch. Sahu, M-Laxmanpur, Bhimpur.	17.60	52500	91-92	212000	2472	247200	12 to 22
8.	Sankarsan Gouda, Laxmanpur, Bhimpur.	5.86	17580	91-92	70500	821	82000	12 to 22

(contd....)

1	2	3	4	5	6	7	8	9
9.	Khalli Das and others, Arunapur Mouza-Bhamarkudi.	2.00	7000	91-92	24000	280	280000	12 to 22
10.	Kama Sethi & others Bhamarakudi	1.00	4000	91-92	12000	140	14000	12 to 22
11.	Bata Pradhan, Arunapur.	8.20	24600	91-92	98500	1148	114000	12 to 22
12.	Damuru Sahu, Arunapur, Bhamarkudi.	1.08	3700	91-92	15000	150	15000	12 to 22
13.	Udaya Sahu, Arunapur. Bhamarkudi.	3.00	9000	91-92	36000	420	42000	12 to 22
14.	Goraya and others Khaganarkodi	9.20	27600	91-92	111000	1288	129000	12 to 22
15.	Kumar Ch. Palai, Arunapur, Bhramarkudi.	5.20	16500	91-92	62500	728	73000	12 to 22
16.	Iswar Naik, Arunapur, Bhramarkudi	3.16	11000	91-92	38000	449	45000	12 to 22

Prawn Farming Technologies

Prawns can be cultured in a number of ways ranging from techniques known as extensive prawn farming through to intensive prawn farming. Extensive culturing techniques generally involve planting low densities of prawn juveniles in earthern dyke or net ponds and harvesting the product in three or four months after the prawns have grown to marketable size. No feeding of the prawns takes place and seedstocks are usually not produced from shrimp seed hatcheries but caught from the wild. Other fish or bird predators are not controlled extensively in this technique. The other extreme of prawn farming is labelled "intensive culturing" and involves the construction of highly engineered earthen ponds, stocking of these ponds to much higher densities, regular feeding of the prawns during their growth cycle, effective control of fish and bird predation, culturing of specific seedstocks, artificial aeration and regular exchange of water. Capital investment in equipment and expertise is much higher with the intensive technique but production per ha. of ponds is also much higher.

A whole gradation of prawn farming approaches also exist, which are generally lumped under the label "semi-intensive culturing". This can range from small operations using almost natural extensive farming techniques but adding some feeding, predation control or artificial seedstock components through to almost intensive farming, where regular ponds are constructed, water is circulated and cultured seedstock used. Production figures of cultured prawns per ha. are used as the criteria for distinguishing between the different culturing techniques. *Penaeus monodon* (tiger prawn) is the species most commonly grown in prawn farming, although other species (e.g. *P. semisulcatus* (banana shrimp) and *P. merguiensis* (white shrimp) are being grown to a greater extent in recent years. *P. monodon* (tiger shrimp) is the species grown in Chilka lagoon by impounding dykes and in net pens.

Pen Culture of Prawns

Pen culture of prawns is demonstrated as a viable one elsewhere in India. A pilot study taken up in Chilka lagoon also yielded a better result. Traditional method of pen culture i.e., simply encircling a particular water area with split bamboo mat and thus allowing the growth of prawns and fish got entangled in that is the way of life for many traditional fishermen even now, which is locally called 'Jano Fishery', which operate during September-October. The harvesting starts from end of November or early part of December when the water level in Chilka slowly goes down. The production of prawns and fish in this method is low and mainly depends upon the density of prawn and fish seeds encircled by bamboo screen.

The penculture trials of prawns in Chilka has been of extensive character. Mostly *Panaeus monodon* has been cultured, but also *P. indicus*. *P. monodon* gets a higher price when sold. In extensive culture the stocked prawns are usually of a bigger size, as the growth is not so fast. It is also common to stock small extensive areas at a little higher density as the expected survival rate is not so high. In smaller pens it will be necessary to feed intensively, while in bigger enclosed water areas feed might not be needed at all as the prawns feed on the natural detritus and bottom-organisms. The availability of feed is already today a bottleneck for an increased production of farmed prawns, at least locally.

Table 10.3: Comparison of the Three Major Methods of Shrimp Culture

	Extensive culture	*Semi-intensive culture*	*Intensive culture*
Pond size	5 hectares or larger	1.8 Ha.	0.25 Ha.
Pond Dykes	Concrete or Earthen	Concrete or Earthern.	Concrete
Stocking Density (PL)	10,000-30,000 per ha.	60,000-1,60,000 per ha.	180,000-300,000 per ha.
Fry Source	Wild	Wild or hatchery.	Hatchery.
Fry Size	PL_{30} -PL_{35} (25-30 day-old postlarvae)	PL_{30} -PL_{35} (20-35 day old post-larvae)	PL_{30} (20 day-old (Postlarvae)
Water management	Tidal exchange, no aeration.	Tidal exchange & pump with aeration.	Pumps, filter, pre-mixed water paddle-wheels for aeration and water circulation.
Feed used	Natural & occasional supplemental feed.	Pelleted Supplemental feed/natural feed.	Formulated complete feed ration.
Culture period	4-6 months	3-4 months	3-4 months
Harvest size	25 pcs/kg.	25-30 pcs/kg.	32 pcs/kg.
Production (Kg./ha./Yr.)	500 or less	3,000-8,000	10,000-20,000
Survival Rate	50% or less	70-80%	70-80%
Crops per year	1-2	2-3	2.5-3.0

In a trial two pens (100 m^2) were used, one stocked with *P. monodon* and the other with *P. indicus*. Production of *P. monodon* was calculated to be 610 kg/ha/2 months. In this trial hardly no feeding was given. In the pens harvest is done by the traditional way of arranging Dhaudi traps and leaderlines. This catching method makes it hard to recover all the prawns. The results on the production of prawns in pens are based upon the stocking with bigger juveniles.

The production by means of adopting a scientific pen culture method can be improved substantially if little extra care is put in for selective stocking of commercially important species, effective predator control, and supplemental feeding. Chilka offers a unique condition, unlike other places, that the construction of pen can be taken up just before start of rainy season in barren lands, making effective predator control. Water level in lakes rises slowly and penetrates through split bamboo mat, thus preventing the entry of fish seeds and other unwanted prawns into the encircled area. Once the water level, inside the pen reaches above 2', commercially important prawn seeds collected from Chilka lake or elsewhere can be stocked. In case, the size of seeds is smaller, a hapa can be erected inside pen for rearing the seeds until it reaches stockable size. Regular feeding with fresh molluscs or artificial pelletized feeds aids faster growth of prawns, as the physico-chemical parameters of water are in good condition. The pen allows free contact of water between the enclosed region and the Chilka lake, thus providing better water quality without any deterioration. This allows high rate of seed stocking which results in higher production through proper fooding unlike in confined pond culture practices.

Cost of Operation

Penculture and other more or less extensive farming activities have a relatively low investment cost and the running cost is not either that high compared with land-based farms with water exchange.

Prawn seed costs about Rs.100 per 1000 seed in most places. The size from the collection centre in Paradeep is 40-60 mm. and if collected locally they are usually a little bigger. Cost for the transport varies between Rs.35-65 per 1000 seed. For locally caught seed of *P. indicus* they pay Rs.80 per 1000 seed.

Budgets for the activity does not really exist but some figures were mentioned by the farmers. For the 4.9 ha. area the estimated yearly expenditure was Rs.38,000 divided into three parts;

Lease value	Rs. 5,000
Seed-cost	Rs.15,000
Labour, feed & fertilizer	Rs.18,000

The estimated income was Rs.1,00,000 per year. For the 12.2 ha. area, the estimated income was Rs.4,00,000 yearly.

It is estimated that 2000 ha. can be developed under pen culture in Chilka apart from the earmarked area of 5005 ha. for pond culture. The capital expenditure involved on pond construction is higher than that of pen and so requires huge finance. In addition, the time required to construct pond takes long period. The above two factors may not pose any problems on pen culture and can be taken up without much difficulty.

Pond culture (confined) of prawns in Chilka periphery has definitely improved the economical status of poor persons as it gives better income than that of agriculture in saline affected areas. However, the total production of prawns from confined pond culture remains at low, when compared to total production. The average yield obtained at the initial stage will continue without any improvement as there is no scope for water exchange. Sometimes, the water quality gets deteriorated, when there is excess application of manures or feeds, thus affecting even the average production level. Much emphasis is to be paid in careful pond management of confined tanks to get the average production year after year. Use of pumps to replenish water in pond may not work due to percolation loss depending upon water level difference between pond and Chilka.

Pen culture of prawns provides substantial scope for increased production through high stocking of seeds and better feeding methods. Capital expenditure is low and the pen construction can be undertaken very fast. Hence, vast areas can be easily covered at a short span and production can be achieved soon.

Conversion of Traditional Fishery Sources to Prawn Culture

Widespread and rampant conversion of various types of traditional fishery sources like "Bahani", "Jano", "Dian" and "Uthapani" has been taken place in Chilka lagoon after the advent of prawn culture system.

The non-fishermen, the primary societies as well as the Government are responsible for such conversion. While Dian and Uthapani fisheries were converted to culture sources by the non-fishermen and Government; other fishery sources like Bahani and Jano areas have been converted both by the Government and some PFCSs either by themselves or through subletting to third parties and also by the encroachers.

Kantei Ghodasipal Nal Jano having an area of 727.49 acres situated in Bahar Chilka was a capture fishery since its inception, serving a passage of flood water of Daya, Bhargavi and Luna rivers and operating as Jano fishery in winter has now declared as prawn culture source by the Collector, Puri (L.No. 1456 dt.21.2.1992). Similarly Panchabatia-Chandabatia Jano, which was earlier used as capture source is now declared as prawn culture source by Puri district administration (L.No.2906 dt.22.4.1992).

Besides, Ac.650.00 of unsurveyed patch of Bahar Chilka adjoining to Kharimuda Jano were directed to be converted to culture sources vide L.No.1494 dt.18.10.1992 of Collector, Puri. In Sairat case No.10 and 20 of 1991-92, Nadiakhia Jano was divided into two parts, Part-1 and Part-2. Part-1 is kept as Jano, but Part-2 has been leased out as culture source to the non-fishermen.

Government of Orissa in the Revenue and Excise Department Order No.CE-Puri-32/85/13758 dt.13.3.1985 have converted Bhobania Jano by alloting Ac.1171.92 in several plots at Panaspada. Ac.582.63 in Plot No.1 and 2 under village Siara has to be converted to culture sources by OMCAD. (Orissa Maritime & Chilka Area Development Corporation), a Govt. of Orissa, Undertaking.

In Mis. case No.11 of 91 the Collector, Puri has directed to divide Ac.581.21 of Jano into culture area, viz. Sarkaria Jayamangala Jano.

The Collector, Puri has directed the Tahasildar, Krushnaprasad to convert Ac.650.00 of Bahar Chilka as stated above be leased out to 30 families of village Sipia under Krushnaprasad Tahasil.

Corporate bodies like OMCAD and Chilka Aquatic Farm are also responsible for conversion of Jano and Bahani areas to culture source, although the latter body seems to have refrained from the adoption of culture fishery in the lake on account of massive public agitation.

Table 10.4

Sl. No.	*Place*	*Previous land position*		*Present position of the land available with OMCAD Corporation*
		Village	*Area in (Ac.)*	
1.	*Panaspada*			
	Puri-Tahasil	Tangarnasi	1171.92	Subleased for prawn culture to the private parties
	Puri-Dist.	Dahikhia	56.48	
		Siara	582.60	
			1810.76	
2.	*Location*	Nandala	417.70	Sub-leased to local villagers for prawn culture.
	Bajrakote	Paikarapur	78.41	
	Krushnaprasad	Talangiri	272.38	
	Tahasil.	Bodakuda	139.32	
	Puri-Dist.	Arakhakuda	131.50	
		Malakuda	233.36	
		Kadakani	153.60	
		Total	1426.27	

206 traditional fishery sources out of a total of 333 have been converted to 248 culture sources as detailed below:

Traditional Sources

	Bahani	Khanda	Jano	Dian	Uthapani	Total
	02	06	82	114	02	206
After conversion to culture sources.	06	06	125	111	2	248

The figures mentioned above reveals that large and unwieldy Jano fishery sources have been split up into culture sources. As such, 82 Janos have been subdivided into 125 culture sources by the government.

Besides, these, the Collector, Puri has reported to have created 61 new culture sources in the fringe areas of the lake between 1988 and 1992.

That the culture source does not impede upon capture source has been belied as many Janos have already been converted to culture sources.

Collector, Puri has reported to have demarcated between the capture and culture fishery sources in Chilka. According to the report, the total number of identified capture sources is 119 with an approximate area of Ac.27019.35 and the number of identified culture source is 314 with an area of Ac.25,068.42. From these figures, it is now clear that as per the 1991 lease policy, the entire fishery sources of the lake has been divided between the capture and culture fisheries at the ratio of 60:40. This means the Government now propose to take away a large chunk of capture sources and get it converted to culture source. The capture fishery sources enjoyed by the traditional fishermen in the lake in the past numbered 203 under the regime of the CFCMS, but now this number will be reduced to 119 alongwith the shrinkage of area. One justification is, however, found in K.P. Biswas Report that as many as 40 per cent of Janos in Chilka has been dried up.

Excavated Ponds in the Marginal Area of Chilka Lagoon for Prawn Culture

Construction of Gheri Bundhs in Chilka Lagoon

Chilka lagoon has lost its scenic beauty today on account of large-scale construction of gheribundhas made either of mud embankment or bamboo, net enclosures for prawn culture. The peripheral and fringe areas all around the lake have been converted to small ponds. Gheribandhas have been constructed in the lake with the onset of prawn culture in eighties. Such gheries have been made by the non-fishermen, fishermen societies, OMCAD, Chilka Aquatic farm and merchants, middlemen and encroachers from outside. They have either put up mud embankments or made net enclosure by fixing bamboo and wooden poles: the latter process is known as "Pen culture". Starting from early eighties, the construction of gheribundhas was accelerated since 1988. It is now estimated that the total areas covered by gheribundhas in Chilka is around 2610.56 ha. as detailed below:

Tahasil-wise break-up of gheri areas in Chilka:

1. Puri sadar	242.75 ha.
2. Krushnaprasad	2367.81 ha.
Total	2610.56 ha.

Table 10.5

Sl. No.	*Name of the Gheries/ Janos and location*	*Area in Ha.*	*Name of the PFCS which availed lease from CMFCS Balugaon*	*Whether direct culture by PFCS or sub-leased to 3rd party*	*Name and address of the sublease holder doing prawn culture*
1	2	3	4	5	6
1.	Raypur Nali (in between Keutakudi & Raypur).	64.00	Gopinathpur PFCS	3rd party	N.Sahoo & others Bramhagiri.
2.	Panda Jano, Panda.	80.00	Jadupur PFCS	-do-	-do-
3.	Siara Gunthali Jano, Siara.	20.00	Villagers of Siara	-do-	-do-
4.	Kaji Bhoisahi Khar, Panaspada.	20.00	Villagers of Panaspada.	-do-	-do-
5.	Dalai Chatara, Panaspada.	16.00	Villagers of Panasapada. R.Pradhan & others.	-do-	-do-
6.	Ganga Pania, Panaspada.	12.00	-do-	-do-	-do-
7.	Lenkasahi Dihan, Panaspada.	4.00	Villagers of Panaspada.	-do-	-do-
8.	Suanpadar Jano Panaspada.	20.00	-do-	-do-	T.Pradhan, Siara.
9.	Sundi Jharia Jano, Panaspada.	40.00	Villagers of Panaspada.	-do-	-do-
10.	Sundi Jharia Jano, Panaspada.	8.00	-do-	-do-	-do-
11.	Dahikhia Nali	28.00	Budhima PFCS, Gangadharpur.	Subleased.	N.Swain & others, Benga Maa.

(contd....)

1	2	3	4	5	6
12.	Nadiakhia Jano, Dahikhia.	28.00	Budhima PFCS, Gangadharpur	Subleased.	S.Dalei & others, Gambhari.
13.	Nuaghari, Nuagaon	10.00	Villagers of Nuagaon.	-do-	N.Prasad & Others, Cuttack.
14.	Pradhan Ghery, Nuagaon.	7.20	Rama Ch.Pradhan.	-do-	-do-
15.	-do-	2.40	-do-	-do-	-do-
16.	Mohapatra Jano, Nuagaon.	2.40	Panchu Mohapatra,	-do-	-do-
17.	Mohapatra Jano, Nuagaon.	2.40	Sankar Mohapatra.	-do-	-do-
18.	Mohapatra Jano, Nuagaon.	2.40	Laxman Mohapatra.	-do-	-do-
19.	Nathapur.	1.60	Braja Mishra	-do-	-do-
20.	Mohapatra Jano.	4.00	Radha Shyama Mohapatra, Nuagaon.	-do-	-do-
21.	Nathpur.	2.00	Lingaraja Mishra	-do-	-do-
22.	Kampana, Gambhari	20.00	Villagers of Gambhari.	-do-	-do-
23.	Gambhari	2.00	-do-	-do-	-do-
24.	Gambhari	6.00	-do-	-do-	-do-
25.	Goseinghar, Gambhari	30.00	-do-	-do-	-do-
26.	Gambhari	5.60	-do-	-do-	-do-
27.	Gambhari	4.40	-do-	-do-	-do-
28.	Gambhari	2.00	Hadu Palei & others.	-do-	-do-
29.	Gambhari	2.00	S.Parida & others	-do-	-do-
30.	Gambhari	6.40	Dwija Dalai & others.	-do-	-do-
31.	Tua	6.40	Golakha Parida & others.	-do-	-do-
32.	Tua	6.40	Kailash Jana & others.	-do-	-do-
33.	Tua	6.40	Nrusingha Parida & others.	-do-	-do-
34.	Gambhari, Badanal	20.00	Villagers of Gambhari.	-do-	-do-
35.	Baghamunda	6.00	Villagers of Baghamunda.	-do-	-do-
36.	Gambhari.	4.00	Sukanta Mansingh & others.	-do-	-do-
37.	Gambhari	20.00	Josi Jena & others	-do-	-do-

(contd....)

1	2	3	4	5	6
38.	Kanhei Jora, Baghamunda.	1.20	Villagers of Baghamunda.	Subleased	N.Prasad & others, Cuttack
39.	Kanhei Jora, Baghamunda.	1.20	-do-	-do-	-do-
40.	Khandia Nala, Baghamunda.	6.00	-do-	-do-	-do-
41.	Sukrabaria, Baghamunda.	4.00	-do-	-do-	-do-
42.	Bahanga, Baghamunda.	6.00	-do-	-do-	-do-
43.	Mahisimunda, Baghamunda.	12.00	-do-	-do-	-do-
44.	Tipalaga, Baghamunda.	16.00	-do-	-do-	-do-
45.	Dalapasada, Baghamunda.	2.00	-do-	-do-	-do-
46.	Damara Nali, Baghamunda.	8.00	-do-	-do-	-do-
47.	Beta Nali, Raypur	4.00	Polangeswari PFCS, Raypur.	M.Swain & others Bramhagiri.	-do-
48.	Sipakuda	20.00	Chaubara, PFCS.	3rd.party	M/s. KMT. Bhubanes-war
49.	Sipakuda	12.00	-do-	3rd party	M/s.Falkon Bhubanes-war
50.	Ghataghari, Golapur	4.00	Villagers of Gambhari.	-do-	-do-
51.	Jagdev Ghery, Gambhari.	14.40	Indramani Jagdev	-do-	-do-
52.	Mansingh Ghery, Gambhari.	14.40	Purna Ch. Mansingh.	-do-	-do-
53.	Gambhari	4.00	Madhusudan Mohanty & others.	-do-	-do-
54.	Gambhari	4.00	Satyapriya Mohanty & others.	-do-	-do-
55.	Gambhari	10.00	Narayan Mohanty, & others	-do-	-do-
56.	Ranaghat, Near Bhabakundaleswar.	3.20	Village Committee, Gabakunda, Mirzapur, Raypur.	-do-	K.C. Jena, Berhampur.
57.	Sanapatna	8.00	-do-	-do-	Tuna Das, Khorda.

(contd....)

1	2	3	4	5	6
58.	Sanapatna.	2.00	Village Committee, Gabakunda, Mirzapur, Raypur	3rd party	Tuna Das, Khorda.
59.	Dharmakuda, Sanapatna.	10.00	-do-	-do-	G. Chelberai, Snapatna.
60.	Kasturanala, Sanpatna.	10.00	-do-	-do-	-do-
61.	Matiakuda, Sanapatna.	10.00	-do-	-do-	Budhia Jena & others.
62.	Dharamakuda, Sanapatna.	6.00	-do-	-do-	Kelu Dalei & others.
			Tahasil-Krushna Prasad Bajrakote Area		
63.	Paikarapur, OMCAD Ghery.	100.00	Belongs to OMCAD.	-do-	Villagers of Paikarapur.
64.	Nandala, OMCAD Ghery.	120.00	-do-	-do-	Nandala.
65.	Kadua Ghery, Nandala.	20.00	Revenue Land	-do-	-do-
66.	OMCAD Ghery, Ramlenka.	130.00	Belongs to OMCAD.	-do-	Villagers of Ramalenka.
67.	Kadua Ghery, Ramlenka	20.00	Revenue Land	-do-	-do-
68.	Siradei Jano, Bramhanideo.	20.00	Revenue Land	-do-	Villagers of Bramhani-deo.
69.	Patanasi Jano, Kandaswar	30.00	Grinath PFCS, Keshipur.	3rd. Party	Villagers of Kandeswar.
70.	Janhikuda Jano, Talatola village.	52.00	Revenue	-do-	Brundaban Pradhan & Villagers.
71.	Arakhar Jano, Talatala Arakhar.	60.00	-do-	-do-	-do-
72.	Nuagaon	54.00	-do-	-do-	Rajkishore Jena & Raj-kilshore Sahoo.
73.	Bramhanikhar, Udayagiri.	15.20	-do-	-do-	Maheswar Dalai and others.
74.	Chatara, Udayagiri.	40.00	-do-	-do-	-do-
75.	Brundaban Sagar, Balarampur.	40.00	-do-	-do-	-do-

(contd....)

1	2	3	4	5	6
76.	Brundaban Sagar, Mundabar.	40.00	Revenue	3rd. Party	Maheswar Dalai and others.
77.	Brundaban Sagar, Hunda.	40.00	-do-	-do-	-do-
78.	Brundaban Sagar, Malud.	10.00	-do-	-do-	Sibajana, Malud.
79.	Ratamati, Brundaban Sagar.	80.00	-do-	-do-	Ratamati, villagers.
80.	Brundaban Sagar, Gopalpur.	80.00	-do-	-do-	Gopalpur, villagers.
81.	Mulakpari, Ratamati.	120.00	-do-	-do-	Villagers of Siandi Ora, Ratamati.
82.	Siala Chatara, Siala.	30.00	Chamunda PFCS, Rambha.	-do-	Bishnu Prasad Das & others.
83.	Bisoikhar, Budhibar.	32.00	Pvt. land of Kishore Bisei & others.	-do-	Purchased by Suryo Udyaga Co. Ltd.
84.	Podakhia, Siandi	249.21	Chamunda PFCS Rambha.	-do-	Subleased to Kanhei Tarai & villagers of Siand.
85.	Chikal Khandi, Talatola.	80.00	Trinath PFCS, Keshipur.	-do-	Subleased to Bihari Pradhan & villagers of Talatala.
86.	Patanasi Jena, Patnasi.	20.00	-do-	-do-	Ghana Behera & further leased to Alia Palei & others of Pitisala village.
87.	Patanasi Jena, Patnasi.	30.00	-do-	-do-	Durga Ch. Baliarsingh of Rasidgaon.
88.	Balighatha, Maleswari.	60.00	-do-	-do-	Raghu Behera, then leased to Kuna Khan of Malud.

(contd....)

1	2	3	4	5	6
89.	Biluamari Jano, Maleswari.	20.00	Trinath PFCS, Keshipur	3rd. Party	Rajkishore Sahoo, Nuagaon
			Bramhagiri Block Tahasil Puri Sadar		
1.	Nua Jano, Machhiadiandi.	10.00	Revenue	-do-	-do-
2.	Khadakhia Jano, Machiadiandi.	12.00	Revenue	-do-	-do-
3.	Singitali Jena, Machhiadiandi.	28.00	Paicha Baral, Parana Math, Narendra Singh of Machhiadiandi.	-do-	Direct cultured.
4.	Badaghori, Haridas.	28.98	Revenue.	-do-	-do-
5.	Haridas Ghery (I)	8.90	Revenue	-do-	-do-
6.	Haridas Ghery (II)	8.90	Revenue	-do-	-do-
7.	Haridas Ghery (III)	8.90	Revenue	-do-	-do-
8.	Haridas Ghery (IV)	8.90	Revenue	-do-	-do-
9.	Haridas Ghery (V)	8.90	Revenue	-do-	-do-
10.	Badakunda, Khotandi.	96.75	Maa Hingulai PFCS, Khatadi.	-do-	-do-
	Total	2610.56			

A gherybundh is made of low height mud wall upto 3 ft. height encircling a large area. As such, a gherybundh is different from a Jano ghery. If Jano gheries are used for capture fishery, gherybundhs are used for culture fishery. Another type of mud embankment has been constructed in Chilka lake under the poverty amelioration programme of FRRP and IRDP. Under these schemes, each beneficiary is provided with Ac.0.50 excavated prawn culture tanks. The excavated earth is used for construction of mud embankment of tank. Besides, the pond prawn culture tanks in the landmass of Chilka lake are also covered by mud embankment.

Out of a total number of 49 PFCSs, 9 societies have made gherybundhs till-date and two societies have proposed to construct the same.

Some PFCSs although did not construct gherybundhs for themselves, yet they are responsible for such large scale construction as they have sublet their fishery sources to third parties for prawn culture. Gherybundhs are not only the necessary outcome of large scale subletting of fishery sources but they are also a necessary evil. They are necessary, no doubt, for prawn culture but there is no gainsaying the fact that they pose a menance to the lake fishery sources and also for its eco-system. The construction of gherybundhs causes hindrance to free flow of water, tide, navigation, free migration of prawn and fish juveniles and also the loss of grazing ground for the juveniles. Besides, the gheries act as silt trap and accelerate the process of silting of marginal areas of the lake.

Table 10.6

Sl. No.	Name of the PFCS	Area under ghery	Year of construction of Gherybundh
1.	Jenamoni PFCS	Ac.126.57	1989
2.	Ambika PFCS	Ac.299.26	1991
3.	Budhima PFCS	Ac.150.00	1991
4.	Ma Mangulai PFCS	Ac.244.41	1992
5.	Chaubardev PFCS	Ac.100.00	1992
6.	Jadupur PFCS	Ac.209.00	1992
7.	Gopinathpur PFCS	Ac.199.00	1992
8.	Jogamaya PFCS	Ac.260.00	1991
9.	Sabulia PFCS	Ac.200.00	1992

Landmass of Chilka Lagoon

Topographically, the Chilka lagoon can be divided into three parts viz. the deeper Chilka otherwise known as 'Bhar Chilka': the shallow Chilka otherwi e known as 'bhittar Chilka' and the landmass. A precise definition of landmass is not possible nor a clear-cut demarcation of the landmass area in the lake is feasible for certain obvious reasons. The numerous variables and parameters governing the extent of landmass in the lake are the rainfall, flood, tides, weed infestation, siltation, wind and storm. The actual area of the landmass may vary from year to year depending on variations in the variables and parameters mentioned above. However, the landmass refers to the portion of the lake, which gets sub-merged under water during the rainy season and dried up during summer.

Roughly speaking, the landmass is measured by the difference between the highest water mark and the lowest water mark in a year. That being so, the difference between the maximum water-spread area of the lake of 1165 sq.km. during the rainy season and the waterspread of 790 sq.km. (ORSAC: Satellite data of May 1986) during the summer constitutes the area of the landmass. The difference of 375 sq. km. (37,500 ha.) may be said to be the landmass area of the lake. But this seems to be an inflated figure as quite a large extent of such areas located in the seasonal dry zone has been converted to agricultural land in the past. Located in the peripheral zone of Chilka, the landmass area remains dry for 7-8 months in a year. More precisely, the landmass comprises Uthapani, Dian and dried up Jano, gramatali areas. As per the record of the Collector, Puri the landmass of Chilka lake adjoining the neighbourhood villages is approximately Ac.9,525 only.

Tahasil-wise break-up of the landmass in Chilka

Banpur	Ac.2047	Approximately
Puri	Ac.2065	Approximately
Krushnaprasad	Ac.5412	Approximately
Total	Ac.9524	

The estimate of landmass, made by the Collector, Puri seems to be a gross under-estimate for the following reasons:

1. In the first place, this estimate does not include the landmass of the lake belonging to Khalikote and Khurda Tahasils.
2. In the absence of accurate revenue records, the measurement of landmass remains doubtful.
3. Most probably, the Collector, Puri might have identified Ac.9524 landmass for culture fishery.

If landmass is the peripheral zone of Chilka, which remains dry for 7-8 months in a year, then ORSAC postmonsoon data of 863 sq.km. waterspread in November, 1985 may be taken as a reliable guide. Now the difference between the maximum waterspread of 1165 sq.km. during the rainy season and the waterspread of 863 sq. km. comes to about 302 sq.km. (30200 ha.). Dividing this area between agricultural land and landmass, the latter may be reasonably estimated at 20 p.c. which comes to about 6040 ha. or Ac.15100. This almost coincides with the relevant data from the Department of Fisheries.

"The macro-level study conducted by the Department of Fisheries, Orissa in the Chilka area showed that 5005.79 ha. brackishwater areas are available for aquaculture, which was earlier estimated at more than 6000 ha. More than 65 per cent of the suitable areas are available in the north-eastern part of the lagoon periphery and the remaining 35 per cent areas are spread over western and south-western part of the lagoon. Most extensive peripheral flats are available between the mouth of river Daya and Dahikhia creek. Brackishwater areas available in the south-eastern part are most undulated in nature (Mohanty, 1988).

Thus the landmass area in Chilka may be taken as Ac.15,100.00.

Aquaculture—A Threat to the Lagoon Environment

The Chilka lagoon is an open ecosystem made up of complex interrelated parts. These components interact, not only amongst themselves, but also, with other major adjoining ecosystems like islands, coastal and bottom vegetation.

Ecological relationships are manifested in physiochemical settings of non-living or abiotic environmental substance and gradients. Included in these groups are basic inorganic elements and compounds, such as calcium and oxygen, water and carbon dioxide, carbonates and phosphates and an array of organic compounds, the byproducts of organic activity. They also include such physical factors and gradients as moisture, winds, currents, tides and solar radiation with its concomitant light and heat. It is against this abiotic back-drop that the biotic components—plants, animals and microbes—interact. A given biotic physiochemical environment and its particular biotic assemblage of plants, animals and microbes constitute an ecological system or ecosystem. Lagoons are highly dynamic, ephermeral acquatic system, which once formed, exist for time periods varying between hundreds and thousands of years. The lagoon lifespan is believed to be positively correlated with the size. "Because of their shallowness, lagoons are relatively easily converted into swamps, marshes, and ultimately land by plant colonisation and encroachment (Except in more saline lagoons). Many of these successional changes are currently operating in Chilka lake.

It is well known from various studies that geophysical and physiochemical changes in the lagoon eco-system will ultimately lead to disappearance of the lake in course of time. The siltation leads to weed coverage and reduces the bathymetry of the lake. A decline in the salinity of the lagoon water will convert the lake into freshwater system, enhancing the weed coverage which holds the silt leading to the emergence of landmass in the fringe areas. The salinity pattern will ultimately determine the fate of closed ecosystem in the Chilka lagoon.

Aquaculture activities will intensify and extend, undoubtedly with environmental implications, either direct or indirect. Once converted into fishponds, they cease to be the nutrient sink of the contiguous coastal zone. This is aggravated by chemicals and excessive organic loads eventually released by these fishponds into the lagoon. Fishpond development will therefore affect, not only those who depend directly on lagoon fisheries, but also the poor families engaged in marginal fisheries of the lagoon.

The lagoon is the foodbase of the complex food web of marine and freshwater organisms including valuable estuarine and diadromous fisheries and ultimately, the fisherman at the top of the food web. Using this area for aquaculture destroys the sanctuary, the nursery and breeding grounds for countless terrestial and aquatic fauna and the nutrient sink that supports the fisheries of contiguous coastal zones, thereby affecting fry abundance and catch by fishermen.

The decrease in lagoon fish catch in recent years attributable to ecological destruction of the lagoon as many diadromous fish are found to graze as juveniles in shallow and marginal areas of the lagoon and destroying this habitat could lead to failure in recruitment.

Advances in Brackishwater Aquaculture

While both intensive and extensive aquaculture development could increase the food production capacity of the aquatic environment, these could also bring about adverse ecological conditions if not properly guided. Once productive jano areas of the lagoon can also be effected by simply disconnecting the areas from adjoining systems by constructing earthen embankments and thereby environmental degradation.

In recent years, especially with the development of sophisticated culture systems for *P. monodon*, stocking densities previously unimaginable are now common. Extensive shrimp culture previously adopted a stocking density of 1-2 per m^2, but new systems now accommodate as many as 50 per m^2.

Needless to say, this stress the pond environment because of metabolites deposited, excess feed (feeding is indispensible at this density) and complications arising from intense intraspecific interactions of a highly crowded shrimp population. Overcrowding results in poor growth of individuals either directly, through deterioration of water quality, or through social mechanisms which govern the overall performance of the population. Individuals weakened by stressful conditions are susceptible to parasites and disease causing organisms which become ubiquitous in highly intensive systems where excessive organic loading provides favourable conditions for their growth.

In the feeding regimes in the culture of *P. monodon*, enormous amounts of food ranging from as low as 0.972 tonnes to as high as 19.44 tonnes per ha. are dumped in the pond over 4-5 months. Since the estimated amount of food converted to shrimp flesh is only 20% on dry weight basis, the remaining 80% takes the forms of faeces, urine and gases (carbon dioxide and ammonia). These excreta provide the major source of nutrients for both pathogenic and nonpathogenic organisms.

In addition, ponds are drained periodically, either for water replacement, preparations or for harvests. Excess material washed out with the water is exported to other systems or back to the original system. The implications to other users of the common water ways become clear. For instance, the widespread incidence of diseases in shrimp farming is attributed to pollution from excessive organic loading in intensive farming. Most of the reported diseases (80%) in farming are caused by protozoans.

Use of Chemicals and Drugs

While the impact of chemical use in aquafarming has not been properly assessed, in due course, accumulated chemical residues will adversely affect not only the industry, but consumers as well. Most alarming is the use of chlorinated hydrocarbons, which are non-biodegradable; although even chemicals, which contain organophosphate as an active ingredient, can also result in fish kills if not allowed to degrade in the pond before being released to adjacent areas.

The following are some chemicals and drugs reportedly used for aquaculture.

Daimetin	Used to purify and sanitize water (absorbs ammonia, carbon dioxide, hydrogen sulfide and other toxic pollutants that retard growth and foster diseases). Active ingredient zeolite (volcanic mineral).
Tan-Pax-So	improves growth of plankton and lablab; complete fertilizer containing N.P.K. and essential trace elements such as Ca, Mg, S, Fe and Boron.
Hai-Chon-Por	algaecide; organic chemical.
F.G.C. Mycin	Shrimp medicine
Endrin	for pests, predators and competitors
DDT	for pests, predators and competitors
Gusathion	for pests, predators and competitors
Brestan	for pests and predators

Malachite green algaecide
Copper sulfate algaecide
Formalin
Potassium permanganate
Bromosept
Flourine

On the question of the impact of the adoption of culture fishery on the ecology of Chilka lagoon, it can be said that Semi-intensive and intensive type of prawn farming will destroy the ecosystem and bio-diversity of Chilka affecting different species of birds, animals and plant life, some of whom belong to the endangered species. The long-term hazardous affects of culture fishery, will include after every crop, which lasts for 2.5 to 3 months, the pens have to be entirely emptied, the polluted mud at the bottom and sides scrapped off. Both the effluent and the sludge affect the ecosystem and the natural balance, but their long-term effect on plant, animal and human life has not yet been possible to determine. The birds may start laying sterile eggs thereby bringing their species to extinction, human beings by consuming fish and prawns enriched in chemicals may show hormonic imbalances. Reports from Taiwan, Thailand, Phillipines, Equador, Malaysia confirm that prawn pens inside lakes and ponds became biologically dead after 3 to 4 years of semi-intensive prawn farming. This is the risk Chilka faces alongwith all other places in our country, where this method of prawn culture is going to be practised.

Illegal encroaching of Janos and other low lying areas of the lagoon by erecting mud-wall and enclosure, not only have a strong ecological impact by cutting off the area from the lagoon eco-system and prohibiting the same to function as nursery and breeding grounds for a variety of organisms, but causing a social conflict between the traditional fishermen of the lagoon and the prawn farmers.

Almost all the fishery scientists and environmentalists have agreed on the point that the adoption of culture fishery in large water bodies like Chilka is positively detrimental to the lake environment and ecosystem. Pen and cage culture in natural water bodies, may have serious impact on the environment, when taken up in a large scale. Hence proper planning should be made before expanding the technology to a greater scale. They have given the following reasons for the possible damage to the eco-system of Chilka lagoon.

(1) Cage and pen structure exert considerable impact on water currents, depending on the flow rate, water density, enclosure size and shape, mesh size, materials and degree of flowing. Significant reduction in flow would cause sedimentation of large denser particles in the immediate vicinity of the cages and pens and thereby disrupting benthic communities.

(2) The low oxygen values in the pen water indicate source of pollution, probably from the decaying vegetation all-round. The lake itself has a dense growth of vegetation, which is constantly disturbed, broken up and even removed by the heavy fishing activity. This result in heavy detrital load. Constant removal of this mass would be needed to create higher or optimal oxygen budget. This feature of low oxygen saturation

has to be critically viewed. The flow of freshwater in monsoon might result in haloclines and is a real critical factor for the pen stock. When such conditions occur, metabolites may start to decompose at the bottom and result in oxygen depletion below critical level.

(3) Enclosure culture systems are actually subcomponent of the lake or estuary in which they are located, and the interactions between the environment inside and the environment outside the enclosure occur with little restriction and so changes in one part of the eco-system invariably affect all other parts to a greater or lesser degree.

(4) Depending on the type of culture employed, the nutrient status of the original ecosystem will be affected. With the traditional system of culture, there will be nutrient loss from the ecosystem through harvested fish and with intensive culture system, where all the requirements of food will be met from outside: there will be nutrient gain in the form of unconsumed food, faecal matter and other wastes. The former will lead to nutrient depletion and the latter will result in organic pollution.

(5) Even extensive prawn culture on the fringes of Chilka is said to have a serious impact on the ecology of the lake. Massive excavations and gherybundhs have been made in such peripheral areas of Chilka, which were originally flooded during the monsoon period. After the conversion of such areas into brackishwater culture ponds, the shallow and grass-covered juvenile nursery grounds have been lost for good and this leads to a sharp decline in the juvenile population. Extensive implementation of brackishwater ponds will result in serious obstructions for the flood water movement and sedimentation rate near the pond embankments leading to eutrophication of the lake. These ponds being completely enclosed in nature are rainfed. With this type of operation, the soil salinity will be leached and the water will become completely fresh after a few years as it happened to Kolleru lake in Andhra Pradesh and Bakhali fish farm in West Bengal. This will render the ponds useless for *P. monodon* culture. Regarding the implementation of pen cage culture either extensive or intensive in Chilka, Rajyalakshmi has warned the fishery scientists and technologists by saying, "Large-scale implementation of this technology should be judiciously undertaken with great care so as not to disturb the existing ecosystem." Comparing the prevailing conditions of Chilka lake with those of Sangkhala lake in Thailand, the CIDA in its report on Chilka mentions: "Knowledge related to shrimp farm development gained at Sangkhala, indicates that intensive or semi-intensive farming should not be carried out on Chilka lake unless conduits for disposing off nutrient-rich effluent directly into the sea are provided.

From every known instance, where prawn culture has been established both the local people and the wet land ecologies have suffered. The strategic course already set by the prawn industry in Taiwan, Thailand, the Phillipines, Equador and many other coastal nations is too evident to ignore.

Between extensive and intensive culture fishery, the latter has much greater hazards on the ecosystem of Chilka than the former. Extensive culture fisheries by way of land-based ponds, Pen culture in the lake will reduce the mixing of lake water with the water either in the pond, cage or in pen enclosure and will thereby create uneven pattern of salinity reducing the output of prawn. Between the two the lesser evil may be accepted.

The lesser evil would be to utilise technology that does not stress the environment in terms of organic and nutrient loading chemical use, and water-power requirements. In contrast Chilka lake in Orissa has to be developed.

Therefore it will be better if aquaculture activities in Chilka lagoon is kept minimum, invoking the Ramsar Convention and citing the social displacement and environmental degradation that are possible consequences.

The planning for commercial prawn farming in the periphery and inner areas of the lagoon, therefore, calls for a multi-disciplinary scientific approach so that the natural prawn fishery of the lake, natural habitat of the resident and migratory birds and above all, the dependence of several thousands of poor fishermen families on the prawn fishery resources of the lagoon remain unaltered.

Economics of Prawn Culture

Scientific prawn farming or aquaculture is an industry and any investment therein is a capital-intensive project, although the capital intensity varies with the type of culture fishery adopted. For example, the capital-intensity of investment in semi-intensive and intensive culture fishery is much higher than in extensive culture system. Table 10.7 shows the capital investment needed for one acre of traditional extensive culture fishery as is being practised to-day in Chilka lagoon. The total cost is divided into average fixed cost and average variable cost. The elements of fixed cost comprise the lease value, the cost of construction of earthen embankment, cost of fishing gears, equipments and crafts, and in case of pen-culture, the cost of synthetic nets and bamboo/wooden poles. The elements of variable cost include the cost of prawn seed/juveniles, the labour cost, the transport cost, the cost of marketing the prawn and the cost of feed and fertiliser.

Table 10.7: Total Cost of Extensive Tide-fed Pond Prawn Culture Per Acre in Chilka 1992-93

		in (Rs.)
	Fixed cost	
1.	Cost of construction of mud-wall gherybundhs	18,660.00
2.	Lease value	800.00
	Variable cost	
3.	Cost of prawn seed/juveniles	}
4.	Labour cost	}
5.	Cost of transport and marketing cost (Input cost)	} 9,500.00
6.	Cost of feed and fertiliser	}
	Total	28,960.00

As indicated above, the total cost of extensive prawn farming per acre of tide-fed pond culture is estimated at Rs.28,960.

The yield of prawn from one acre of pond culture is estimated at 400 kg. per annum, which is valued at Rs.80,000. The net income from one acre of prawn culturing comes to about Rs.80,000 – Rs.29,000 = Rs.51,000

The cost and benefit from prawn culture as has been computed above reflect only the private cost and private benefit. Estimate of viability of prawn culture project on the basis of private cost and benefit does not truly reflect the economics of prawn culturing in a lagoon like Chilka as it involves significant elements of social cost. Hence the viability of such a project requires a careful preparation of a matrix containing the elements of private cost and social cost on the one hand and elements of private benefit and social benefit on the other. The full social cost of prawn culture includes the cost of production plus the 'environmental' cost, 'displacement cost' and 'user cost'. 'Environmental cost' is the cost imposed on society through any damage to the eco-system and other resources as a result of degradation and pollution. 'Displacement cost' refers to the cost imposed on the traditional fishermen on account of conversion of their erstwhile traditional fishery sources to culture sources. 'User cost' is the value of future resource benefits foregone because they have been reduced by current use. Similarly, the elements of social benefit from prawn culture may include the additional foreign exchange earning accrued to the nation giving a balance of payments support increase in the State Domestic Product and income and net increase in employment, if any. Private cost and benefit are to be weighed against social cost and social benefit in order to arrive at a correct economic estimation of prawn culture projects.

Prawn Trade of Chilka Lagoon — Market Price & Export

Prawn is an important item of export in Orissa's inter-regional trade and it is also a foreign exchange earner as a sizeable portion of the prawn catch from the State goes to the export market destined for foreign shores every year. Prawns are of different size and are of varying types, e.g. Tiger prawn otherwise known as *Penaeus monodon*, white prawn known as *Penaeus indicus* and Metapenus and others. The other classification is penaeid prawn, medium prawn and small prawn. The market price of prawn varies with the size and type from Rs.20 to Rs.350 per kg. The price of export brand *Penaeus monodon* (tiger prawn) is the highest and that of small shrimps is the lowest. Table 10.8 shows the price of standard 20 counts Penaeus monodon per kg. in December 1990, 1991 and 1992 both at the source and at the market. The price at the source is the catcher's price and the market price is the price for the traders.

The size of prawn varies between 10 counts to 40 counts. e.g. 10 nos/ per kg. 20 nos/kg., 30 nos/kg. and 40 nos./kg.

Table 10.9 shows the total prawn production in Orissa and its distribution between inter-state export, foreign export and domestic consumption from 1988-89 to 1992-93.*

---

* It may be noted here that the difference between col-2 and col-5 constitutes the domestic consumption of prawn in Orissa.

Table 10.8: Price of Prawn Both at the Source and in the Market, 1988-92

Year	*Price at the source*	*Price in the market (in Rs. per Kg.)*
1988	180	250
1990	230	300
1991	250	320
1992	280	350

Source: Directorate of Fisheries, Orissa.

Table 10.9: Prawn Production and Export from Orissa: 1988-89 to 1992-93. Year-Total Production Inter-State Export

Year	*Total production (MT)*	*Inter-State export (MT)*	*Export to foreign countries (MT)*	*Total export (MT)*	*Value of inter-state export (Rs. in lakhs)*	*Value of foreign export (Rs. in lakhs)*	*Total value of all export (Rs. in lakhs)*
1988-89	6178	662	2581	3243	198.60	3156.11	3354.71
1989-90	6634	864	3740	4604	259.17	3644.54	3903.74
1990-91	7713	541	3841	4382	189.35	5352.80	5542.13
1991-92	8905	971	3685	4656	339.85	6747.81	7087.66
1992-93	10377	967	4705	5672	386.80	9014.99	9402.79

Source: Directorate of Fisheries, Orissa.

Table 10.9 shows that Orissa's export of prawn increased from 3243 M. Tonnes in 1988-89 to 5672 M. Tonnes in 1992-93: more than 70 per cent increase in a period of 5 years and this has happened so largely for the expansion of prawn production through culture sources. Increase in the quantum of export coupled with a rise in the price of prawn, both taken together have contributed slightly to less than three times increase in the value between 1988 and 1992. One striking feature in Orissa's prawn export during this period is the preponderance of export to foreign countries vis-a-vis domestic export and as such, prawn is a major foreign exchange earner.

Table 10.10 gives a picture of export of prawn from Chilka lake from 1988-89 to 1991-92.

As compared with 1989-90, Table 10.10 indicates that there is a decline in the production of prawn both from capture and culture sources in Chilka in 1991-92. But this decrease in production has been more than compensated by a rise in the price resulting in a significant increase in the value of export. That means the barter terms of trade for Orissa's prawn have been more favourable in 1991-92 that what they were in 1988-89.

Table 10.10: Export of Prawn from Chilka lake: 1988-89 to 1991-92

Year	*Capture from Chilka lake (MT)*	*Culture in Periphery (MT)*	*Total Qty. exported (MT)*	*Total value (Rs. in lakhs)*
1988-89	240.3	375.6	615.9	753.2
1989-90	343.0	628.7	971.7	947.4
1990-91	140.6	590.0	731.0	1020.0
1991-92	231.0	469.0	700.0	1400.0

Source: Directorate of Fisheries, Orissa.

The following enterprises have been listed as the registered exporters of Orissa's prawn. As many as 21 firms are engaged in prawn trade in Orissa.

1. M/s. Suryo Udyog Ltd.
2. M/s. Aditya Udyog
3. M/s. Suryo Foods & Industry Ltd.
4. M/s. Falcon Marine Exports Ltd.
5. M/s. Trimurty Exports.
6. M/s. Elljee Exports.
7. M/s. Raj Exports.
8. M/s. Marine Exports.
9. M/s. Shrimps Ltd.
10. M/s. G.P. Exports.
11. M/s. East Coast Exports Pvt. Ltd.
12. M/s. Coastal Exports
13. M/s. Basanta Foods Ltd.
14. M/s. Amalgan Foods Ltd.
15. M/s. Orissa Shrimps.
16. M/s. Oceanic Exports
17. M/s. Kayee Exports.
18. M/s. Tulip Exports Ltd.
19. M/s G.F. Kellner & Co.
20. M/s. Triveni Food Products.
21. M/s. TISCO Ltd.

11

Conflicts and Problems

Traditional Right of the Non-fishermen for Fishing in Chilka

During the pre-vesting period both fishermen communities and non-fishermen in groups used to obtain the fishery sources by paying royalty or license fee to the Raja of Parikud in accordance with the traditional practice. After vesting in 1953, when the Chilka fishery sources were auctioned publicly by the Anchal Adhikari both the fishermen and non-fishermen communities participated in the auction and this practice continued up to 1959. In the Chilka Reorganisation Scheme of 1959, the open auction was suspended and all the fishery sources vested in the Government in Revenue Department were leased out to the Central Fishermen Co-operative Marketing Society (CFCMS) and this Apex body used to sub-lease fishery sources except Dians to the Primary Fishermen Co-operative Societies. 88 out of 114 Dian fisheries were settled in favour of individual non-fishermen and 30 non-fishermen villages by the Central Society.

Concessions were given to the non-fishermen by way of poluha right in Jano area from 1st March to 31st. July and right to angling in the lake for purposes of their consumption only. Any catch in excess of their consumption was required to be surrendered to the nearest PFCS at the current market rate. According to this stipulation the same Jano fishery sources were used both by the fishermen and non-fishermen communities in different periods in a year. The former enjoy it from 1st August to the end of February and the latter from 1st March to 31st July in a year.

The non-fishermen did not enjoy the traditional right to fishing in Chilka in the same way as their fishermen counterparts did. However, they had enjoyed a traditional right to fishing in Chilka which was confined mostly to Dian fisheries, Uthapani and Chulimunda. The lease records of the CFCMS reveal that the non-fishermen were sometime leased out limited Jano and prawn khandas by the Central Society. They have prayed for recognition of their fishing right time and again, but Government have turned down their claim for fishery rights except to give them certain concessions to fishing in the lake for consumption purposes in 1967. However, a partial recognition of the fishing right of the non-fishermen by the Government seems to have taken place in 1981, when government have said that the persons eligible to be the members of fishery Co-operatives must be either fishermen by caste or by profession or persons exclusively depending upon pisciculture as a profession.

The CIDA in its Chilka lake project report has observed in this regard: "The land belongs to the Revenue Department and the traditional fishermen are the lease holders to fish and hence only the traditional fishermen have the sole right over the lake water. This they have enjoyed from time immemorial and they seem to have documents to this effect from the days of the Maharajas. Caste-Hindus (meaning the non-fishermen) have forcibly occupied the land and water or have taken the water rights on a sub-lease basis. The caste-Hindus use their social, economic and/or mere physical muscle power to retain hold over the land and water and also to acquire more land. The traditional fishermen are increasingly realising this social and economic sub-ordination and want to regain control over their traditional rights and to change this subordinative status. Village level opposition to this subordination and efforts at regaining control over their traditional rights have led to physical conflicts and to law and order problems. There have been continuous occurrences of blood-shed over these issues. The unequal competition between the traditional fisherman and others to gain control over resources will only increase as fishing becomes more profitable and as more and more people with capital to invest are allowed to enter the area.

The caste-class-occupation relationships have undergone a radical change in modern times. Occupations are no longer caste or class-based. There exists a caste-cum-class distinction between the fishermen and non-fishermen. The latter belong to superior caste and class and the former belong to socially inferior caste. So long as fishing in the lake was considered to be an inferior job, it remained as the exclusive domain of the fishermen. Consequent upon the advent of culture fishery in Chilka, the situation has dramatically changed and the non-fishermen have taken to fishing in the lake by giving up all social and psychological inhibitions, which they had in the past. Today the occupations are no longer caste-based.

Non-fishermen Taken Up Fishing in Chilka

Many non-fishermen in the neighbouring villages of Chilka have taken to fishing as their profession since long. It is difficult to ascertain the exact number. In the past they used to fish either through the help of the fishermen or used to catch fish in Janos, Dians, Uthapanis for themselves through poluhas and Khainchis during a specific period, when Janos cease to operate. With the onset of culture fisheries in the eighties, most of the non-fishermen are engaged in fishing. It is reported that the total number of non-fishermen engaged in fishing in Chilka in 1992-93 is 71,244 as shown in Table 11.1.

Table 11.1: Tahasil-wise Break-up of the Number of Non-fishermen Engaged in Fishing Operation in Chilka

1.	Banpur	34 villages		13,320
2.	Puri	45 villages		24,000
3.	Krushnaprasad	75 villages		33,924
			Total	71,244

Sub-leasing of Fishery Sources to Non-fishermen

Till 1988 the Government did not demarcate the various fishery sources in Chilka. In the absence of detailed guidelines the Primary Societies experience difficulties in operating the sources. "The Collectors, Puri and Ganjam were vested with the responsibility of demarcation of various fishery sources in Chilka within their respective jurisdiction. A survey of the lake fishery sources was undertaken by the Collector Puri between 1987-88. But the survey of Chilka fisheries of 1988 is highly faulty as there exists many unsurveyed patches in the lake even today, and disputes persist among the fishermen and non-fishermen over the question of boundaries of different sources in the absence of exact delimitations. In some cases there are discrepancies between the recorded areas and the areas under the actual possession of fishermen/non-fishermen, which obviously refer to the unsurveyed patches.

Some fishery sources in the lake with discrepancies between the lease area and the area under actual possession are the Kerandia Jano, Panchakuda Jano, 42-Bahar Chilka, Ghodabat, Boroganasi, Kathpentha, Ganeśwarnasi, Barunakuda-2, Mauli Boroga and Jagadahan prawn Khandas. It is interesting to note that the C.F.C.M.S. has sub-leased all the above mentioned prawn khandas beginning from Ghodabat to Jagadahan measuring a recorded area of Ac.314.22 cents in 1988-89, to Durga P.F.C.S. as against the actual possession of nearly Ac.2000. There are also cases where the lease area falls short of the recorded area.

The State Govt. also stipulates that all fishery sources of Chilka lake shall be leased out in favour of the C.F.C.M.S. and this Apex body will sublease these sources to affiliated PFCS for a period of 3 years beginning from 1988-89 to 1990-91. Accordingly, the CFCMS has settled a total number of 203 fishery sources in favour of 49 PFCSs in 1988 for a period of three years as detailed below.*

Jano	83
Prawn	70
Bahani	42
Dian	4
Uthapani	4
Total	203

Besides, these, the C.F.C.M.S. has also given lease of 84 Dian and 8 Jano fishery sources to the non-fishermen villages in 1988-89 for a period of 3 years.

The most remarkable change in the lease policy of 1988 is a 3-year period lease operation as against annual lease in the past. Clause-12 of the Lease Agreement of CFCMS stipulates: "The sub-lease or the sub-lessees as the case may be are required to operate the fisheries by themselves and have no power to sublet further. In the event of such subletting, the sublease will forfiet the right of sublease and the lessee shall be competent to enter into and take over the possession of the fishery source. Neither the lessee nor the sub-lessee shall be entitled to claim compensation from the lessor in such circumstances. "This condition has hardly been implemented by the Central Society and a number of PFCSs have reportedly sublet their fishery sources to third parties, e.g. either non-fishermen or merchants and middlemen, as discovered below.

* The record of C.F.C.M.S. shows that although 58 PFCSs were affiliated to the C.F.C.M.S. in 1988, 49 PFCSs were given the lease and the rest were liquidated and defunct.

1. Narayani PFCS has sublet about Ac.100 from Uparachatra Jano at Khatiakudi to one Ganeswar Das of Puri for prawn culture for a period of two years i.e. 1990-91 and 1991-92. The Ganeswar Das has encroached another unsurveyed patch of Ac.200, which is contiguous to Uparachatra Jano. Interestingly enough, Ganeswar Das has again sublet the total area of Ac.300 to one Hazari Palla of Badakhatiakuda for a period of two years, 1990-91 and 1991-92. This is an example of two-tier subletting of the lake fishery source.

2. The Kanak Durga PFCS has sublet Ac.99.20 from Deula Jano for prawn culture.

3. An agreement has been made by the villagers of Badakuda-cum-members of Jenamani PFCS dt.10.3.1990 to the effect of subletting of 1350 ft. of Kathapentha Jano of one Chaubar Dalai of village Anlakuda.

4. One Ramakrushna Das has taken the sublease of Sankha Jano from Lakshmi PFCS and again transferred it to Prafulla Kumar Jena and Dhirendra Jena for a period of 3 years from 15.10.1992. This is also a case of two-tier subletting.

5. One Arkhita Behera, on behalf the fishermen villagers of Badakuta, has sublet a portion of the Godabata Jano to the non-fishermen villagers of village Gurubai. One Nakula Behera of village Badakuta has sublet another portion of the same Godabata Jano to one Jayakrushna of village Gurubai on 18.12.1989. The said Godabata Jano is the leased fishery source of Jenamoni PFCS.

6. The Fishermen villagers of Raskudi are alleged to have sublet a creek or Nala in the lake to one non-fishermen in the name of Biswanath Pradhan of village Jagirkuda.

7. One fishermen in the name of Jaganath Mallick has made an agreement on behalf of the fishermen villagers of Khatiakudi with a non-fishermen in the name Sankar Pradhan of village Prasimpur. The villagers of Khatiakudi being members of Trinathdev PFCS have sublet Raskudi Jano to Sankar Pradhan.

8. Members of Jagulei PFCS have complained that the non-fishermen villagers of Gambhari, Poradi and Gola have enroached the following Janos, which legally belong to them. They took the lease in respect of those Janos from the CFCMS in 1988. These Janos are:

(a)	Barapania Jano	680.00 Ac.
(b)	Panchadera Jano	275.00 Ac.
(c)	Nuakhar Majhi Jano	475.00 Ac.

9. Persons in the name of Jogindra Behera, Krushna Dakua and Hadu Das have, on behalf of the villagers of Biripadar, signed an agreement with one Harekrushna Jena to the effect of subletting Ac.8.00 from Dollamunda to Deula Bada Muhan for Rs. 16,000.00 on 15.6. 1989 for prawn culture for a period of three years from 10.6.1990 to 14.2.1993.

10. Kartika Das, Dasarathi Das, Narasingha Karji, Maheswar Karji, Laxmidhar Behera, and Dandanasi Behera of Biripadar village have sublet Ac. 8.00 to Harekrushna Jena on 19.7.1987 for Rs. 10,000.00 for prawn culture for a period of two years from 19.7.87 to 19.7.89.

It reveals from the Collector's report that as many as 42 PFCSs out of a total number of 49 have sublet their fishery sources to third parties between 1988-92.

Table 11.2

Sl. No.	*Name of the P.F.C.S. (Lease holders)*	*Lease out by C.F.C.M.S. Balugaon*	*Name of the Fisheries*	*Extent of area (Acre)*	*Subleased to (third parties)*
1	2	3	4	5	6
1.	Narayani PFCS, Balugaon (Pathara)	C.F.C.M.S., Balugaon	Salakudi & Chhodara	445.96 & 56.66	Villagers of Sana-Anlo
2.	Mahamaya PFCS, Balugaon (Langeleswar)	C.F.C.M.S., Balugaon	Pandara Jano	185.00	Villagers of Ipping Bankinalli (Faghu Nath)
3.	Narayani PFCS, Balugaon	C.F.C.M.S., Balugaon	Upper chatara	94.15	Villagers of Pathara and Ganesh Das of Langeleswar.
4.	Narayani PFCS, Balugaon	C.F.C.M.S., Balugaon	Talachatara	114.45	Villagers of Khatiakudi and Ganesh Das of Pathara.
5.	Trinathdev P.F.C.S.	C.F.C.M.S., Balugaon	Rasakudi Jano	382.29	Binod Sabu, Balugaon.
6.	Trinathdev P.F.C.S.	C.F.C.M.S., Balugaon	Matia Jano	475.30	Villagers of Bada-Anla.
7.	Kholamuhan P.F.C.S.	C.F.C.M.S., Balugaon	Tanpadakana Jano	130.45	Villagers of Kanadaragaon & Rasakudi.
8.	Alandapatna P.F.C.S.	C.F.C.M.S., Balugaon	Ambrutanidhi Jano	593.10	Villagers of Anua.
9.	Villagers of Titipa	C.F.C.M.S., Balugaon	Bhuruti Jano	441.88	1. Laxman Pradhan of Sana-Anlo. 2. Mahendra Jena to Titipa
10.	Mahamaya P.F.C.S.	C.F.C.M.S., Balugaon	Jayamalla Jano	677.00	Villagers of Pichhini.
11.	Mahamaya P.F.C.S.	C.F.C.M.S., Balugaon	Gorasara Jano	86.35	1. Padma Charan Jena of Krushnaprasad. 2. Trinath Balabanta Roy of Sikharpadar. 3. Basanta Nayak of Chhotapadar.

(contd....)

Sl. No.	Name of the P.F.C.S. (Lease holders)	Lease out by C.F.C.M.S. Balugaon	Name of the Fisheries	Extent of area. (Acre)	Subleased to (third parties)
12.	Mahamaya, P.F.C.S.	C.F.C.M.S., Balugaon	Chikili Jano	35.50	1. Padma Charan Jena of Krushnaprasad. 2. Trinath Balabanta Roy of Sikharpadar. 3. Basanta Nayak of Chhotapadar.
13.	Taradevi, P.F.C.S.	C.F.C.M.S., Balugaon	Khanimuhan Jano	1002.72	1. Umakanta Mishra, Bhubaneswar. 2. Bhaja Das, Puri. 3. Tuna Das, Puri. 4. Indra Palei, Sipia 5. Ekadasi Palei, Sipia.
14.	Mahamaya, P.F.C.S.	C.F.C.M.S., Balugaon	Panchakudi Jano	400.00	Jagannath Pattanaik, Supervisor, Tulia Export Coop., Balugaon, Puri.
15.	Jenamani, P.F.C.S., Badkul	C.F.C.M.S., Balugaon	Godabata Jano	52.15	Villagers of Nuapada & Gurubai.
16.	Nairi P.F.C.S., Balugaon	C.F.C.M.S., Balugaon	Sankha Jano	604.16	About Ac.60.00 to villagers of Analakuda & others.
17.	Harachandi P.F.C.S., Nuapada	C.F.C.M.S., Balugaon	Boraga Jano	98.74	Nalinikanta Jena of Nuapada
18.	Harachandi P.F.C.S., Nuapada	C.F.C.M.S., Balugaon	Jagadhan Jano	273.52	Maguni Charan Pradhan of Nuapada.
19.	Kanakdurga P.F.C.S., Biripadar	C.F.C.M.S., Balugaon	Kantei Jano	321.96	Villagers of Badajhad.
20.	Kanakdurga P.F.C.S., Khirisahi	C.F.C.M.S., Balugaon	Deula Jano	288.16	Villagers of Nuagaon Gonundia.
21.	Kanakdurga P.F.C.S., Khirisahi	C.F.C.M.S., Balugaon	Akashzoda Jano	110.00	Gangadhar Jena of Berhampur.

(contd....)

Sl. No.	Name of the P.F.C.S. (Lease holders)	Lease out by C.F.C.M.S. Balugaon	Name of the Fisheries	Extent of area. (Acre)	Subleased to (third parties)
22.	Jenamani P.F.C.S., Barkul	C.F.C.M.S., Balugaon	Kudakudi Jano	74.42	Chaubar Dalai of Anlakuda & some persons of Baghamunda.
23.	Khatisahi P.F.C.S.	C.F.C.M.S., Balugaon	Badajore Jano	310.62	Basudev Pradhan of Nuapada.
24.	Khatisahi P.F.C.S.	C.F.C.M.S., Balugaon	Sanajore Jano	131.28	Basudev Pradhan of Nuapada.
25.	Aundala P.F.C.S.	C.F.C.M.S., Balugaon	Dhanahara Guachori Jano	189.88	Villagers of Jahnikuda & Lessee himself.
26.	Berhampur P.F.C.S.	C.F.C.M.S., Balugaon	Kolikuda Jano	53.30	A person of Titipa.
27.	Nairi P.F.C.S.	C.F.C.M.S., Balugaon	Sampana Jano	235.54	Tuna Das of Puri.
28.	Keshpur P.F.C.S.	C.F.C.M.S., Balugao.	Chikalkhandi Jano	2375.00	Villagers of Diankuda.
29.	Ambika P.F.C.S.	C.F.C.M.S., Balugaon	Aendakhai Jano	299.26	(1) Bhaja Das of Puri, (2) Tuna Das of Puri.
30.	Trinathdev P.F.C.S.	C.F.C.M.S., Balugaon	Bhagat Jano	69.65	Villagers of Dinakuda.
31.	Trinathdev P.F.C.S.	C.F.C.M.S., Balugaon	Hota Jano	78.33	Villagers of Budhibar.
32.	Trinathdev P.F.C.S.	C.F.C.M.S., Balugaon	Baligotha Jano	151.18	Villagers of Sahanajpur.
33.	Trinathdev P.F.C.S.	C.F.C.M.S., Balugaon	Jalumuhan-Boiluzai Jano	215.38	Villagers of Kaleswari.
34.	Gauranga Behera	C.F.C.M.S., Balugaon	Pedakhia Jano	723.28	Villagers of Standi.
35.	Jageswari P.F.C.S., Alupatna		Dandihida Jano	40.00	Prafulla Jena & others of Satapada.
36.	Budhima, P.F.C.S., Gangadharpur	C.F.C.M.S Balugaon.	Nadiakhia Jano	110.70	Arakhita Jena & others of village Gambhari.
37.	Jadupur P.F.C.S., Jadupur	C.F.C.M.S Balugaon.	Debakha Jano	124.14	Satya Pradhan & others.
			Upargargadia Jano	55.00	Villagers of Bhubanpur.

(contd....)

Sl. No.	Name of the P.F.C.S. (Lease holders)	Lease out by C.F.C.M.S. Balugaon	Name of the Fisheries	Extent of area (Acre)	Subleased to (third parties)
38.	Jogamaya PFCS, Koriapur, Mudiratha & Keutakudi.	C.F.C.M.S Balugaon.	Kerandia Jano	40.00	Panua Sahu & others of village Mudiratha
39.	Gopinathpur PFCS, Gopinathpur	C.F.C.M.S Balugaon.	Asipur Jali Jano	29.05	Parikhita Delai & Sankar Pradhan of village Brahmagiri.
			Kanteigouda Sipela	257.90	Prasana Parida & others of village Rodanga.
40.	Chaubaradev PFCS, Sipakudapatna	C.F.C.M.S Balugaon.	Chandamaranali Bahani	92.00	M/s. Kalinga Marine Transport Co. Ltd., Bhubaneswar.
41.	Mangala PFCS, Barakudi & Baghalanji	C.F.C.M.S Balugaon.	Kodibatia Jano	88.20	Chitaranjan Jena of Oukudi.
				10.00	Ranjan Khan of Mudiratha.
42.	Bagdevi PFCS, Stara	C.F.C.M.S Balugaon.	Bhobania Jano	240.00	Srinath Samantray of Balisahi.

Thus both the field evidence and Collector's report taken together amply suggest a rampant subletting of the leased-out fishery sources in Chilka lake.

The plausible seasons for subletting of fishery sources by some PFCSs are as follows:

1. The Central society has no manpower to detect and enforce violation of the lease terms. No such violation has ever been detected nor any PFCS has ever been taken to task for this.

2. The onset of culture fishery in the lake in the eighties has provided a fillip to subletting. Culture fishery requires heavy capital investment, which neither the societies nor the poor fishermen could afford to undertake for themselves. Alternatively, they were given temptations by the non-fishermen and merchants to part with their fishery rights for an exorbitant sum of money. Those societies which have made such clandestine deals have made a profit to be distributed among its members. The profit is accrued due to the difference between the subletting fees charged on the one hand and the lease value payable by the concerned PFCSs to CFCMS in respect of particular fishery sources on the other.

3. The age-old and traditional indebtedness of the fishermen community to merchants, middle-men and money-lenders is another reason for subletting.

4. Above all, the 3-year lease operation in 1988 seems to have provided an additional opportunity for the PFCSs to sublet their fishery sources.

5. The uneven and erratic distribution of fishery sources by the CMCFS among the various PFCSs is another contributory factor to subletting. For example, primary societies like Kanak Durga and Trinathdev and Mahamaya were given large number of fishery sources as against Bagdevi, Arakhakuda and Chilka Jaganathpur which were given a few sources each. Besides, there are some unwildy Jano areas, which are contiguous to Dian fisheries. Jano areas are being frequently tresspassed by the non-fishermen. Finding this situation of a large number of fishery sources, unwildy size of Janos and unauthorised catch, these societies as well as their members found it profitable and convenient to sublet to outsiders for prawn culture and thereby they earn an income without any labour.

Biswas (1989) reports about the subletting of Jano fishery sources in the lake by some PFCSs, "Because of high prices for prawns, unauthorised conversion of a fishery system creates conflicting interest in utilising the Chilka lagoon in the recent years. Prawn farming now has been taken up in janos and in other enclosed areas of the lagoon, which are shallow and situated close to the land by entrepreneurs illegally. These small janos (4.9-12.2 ha.) are not used by the primary fishermen co-operative societies, even though they take them on lease, but sublease the areas to the interested private entrepreneurs at a higher value."

12

Infrastructure Facilities

Fish landing jetties to facilitate landing of fish catch from the lagoon have been constructed in the periphery of Chilka lagoon as shown in Table 12.1.

Table 12.1

Name of the fishery jetty & location	*Year of completion*	*Cost in Rs.*
1. Nairi jetty	1989-90	6,38,000
2. Balugaon jetty	1979-80	2,66,900
3. Kaluparaghat jetty	1979-80	3,08,370
4. Pathara jetty	1989-90	4,12,000
5. Sabelia jetty	1989-90	2,63,000

There are twelve landing centres around the shores of the lagoon for landing and assembling catch of Chilka fish and prawn as follows:

1. Balugaon
2. Arakhakuda.
3. Nairi
4. Kaluparaghat
5. Bhusandapur
6. Soran
7. Boradi
8. Mangalajodi
9. Rambha
10. Sabelia
11. Pathara
12. Keshpur

Eight processing plants to process the prawn (freezing) of the lagoon have been established on the shores of the lagoon with their capacity as shown in Table 12.2.

Sixteen ice plants, established on the shore of Chilka for preservation and transport of the lagoon fish catch are shown in Table 12.3 indicating their location and ice producing capacity.

Table 12.2

Name of the Plant	*Location*	*Processing capacity in tonnes*	*Cold storage capacity in tonnes*
M/s. G.F. Kellner & Co.	Brahmagiri, Puri	4.0	60
M/s. P & S. Pvt. Ltd.	Brahmagiri, Puri	10.0	100
M/s. CIFOODS Ltd.	Pathara, Ganjam	—	40
M/s. George Maijo Exporters Pvt. Ltd.	Balugaon, Puri	7.5	80
M/s. Mathoor Marines	Balugaon, Puri	—	50
M/s. Orissa Marine Industries Pvt. Ltd.	Balugaon, Puri	3.0	50
M/s. Surya Marine Pvt. Ltd.	Kalupadaghat, Puri	3.5	60
M/s. Kalinga Sea Food Pvt. Ltd.	Kalupadaghat, Puri	1.0	35

Table 12.3

	Name of the Plant with Location	*Capacity of producing ice (in tonnes)*
1.	M/s. Tirupti Ice Factory, At-Keshpur Junction, Khallikote	10.00
2.	M/s. Chilka Lake Ice Factory, Banpur Road, Balugaon	20.00
3.	M/s. Bhima Ice Factory, Kalupadaghat	15.00
4.	Govt. Ice Factory, Chilka Road, Balugaon	5.00
5.	Central Coop. Marketing Society, Chilka Road, Balugaon	5.00
6.	George Maijo and Co., Chilka Road, Balugaon	20.00
7.	Jaihind Ice Factory, Balugaon	10.00
8.	Orissa Marine Industries (P) Ltd., Utarayani, Balugaon	10.00
9.	Jayabala Ice Factory, Panaspada, Puri	20.00
10.	Jaydurga Ice Factory, Sorana	20.00
11.	Sea Lords & Sea Foods, Panaspada, Puri	10.00
12.	Cifood, Pathara	10.00
13.	Ugratara Ice Factory, Sorana, Kuhudi, Puri	10.00
14.	Maruti Ice Factory, Nairi, Puri	10.00
15.	Kalupada Ice Factory, Kalupadaghat	10.00
16.	Chilka Ice Factory, Kalupadaghat	10.00

13

Conservation and Development Proposals

Potential Impacts of Shrimp Farming

Chilka, a coastal tropical lagoon is currently in transition towards a progressive reduction in salinity. This is evident in reduced average salinities over period of years. Besides there has been an observed increase in freshwater flora and fauna in the northern section of the lagoon. This transition will affect the longterm viability of aquaculture operations as they exist in the lake today. Altering the hydrology of the lagoon by mechanical changes (widening and deepening the lagoon mouth and outer channel by dredging) may arrest this transition to the point of reverting freshwater areas of the lake back to more saline (brackishwater) conditions.

Chilka lagoon occupies an approximate area of 900-1200 sq. km. with average depth between 1.0 and 1.3m and water temperature ranging 23° to 33 °C has salinity levels ranging from fresh to saltwater, depending on location and season. During the dry season (February-September) the average salinity is more than 20 ppt, but during the rainy season (October-January) salinity decreases below 20 ppt. In northern portions of the lake freshwater conditions are encountered during the rainy season. Within four kilometers of the mouth sea water conditions persist for most of the year.

Dissolved Oxygen levels generally lie between 18 ppm, although very low levels (i.e. 1.5 ppm) have been recorded in the northern areas due to high BOD (Biological Oxygen Demand) and COD (Chemical Oxygen Demand) leading from swampy areas.

From time immemorial, the lagoon has supported a subsistehce and commercial fishery, provided a means of transportation and most of the population depends on the in-lake and coastal fishery as major and minor occupations.

During last decade, the lagoon has come under increasing use pressure by an expanding local population. Nutrient loading of the lake is increasing caused by domestic waste, agriculture and aquaculture practices. Sedimentation rates in the upper and more sheltered areas have increased.

Residents have noticed reduced fish catches, deteriorating water quality and increasing problems associated with increasing freshwater conditions.

The annual fishery yield of the lagoon is deteriorating to be 3329 tonnes (1993-94). Overfishing deteriorating water quality associated with increasing nutrient loads from domestic waste and run-off from agricultural lands are probably contributing factors to a declining fishery.

While the opportunity for aquaculture seems tremendous there are environmental limitations to further expansion. Initial results indicate that shrimp farm effluents contribute to increased sediment (TSS) and nutrient loadings,

Harvest of Tiger Prawn (*P. monodon*) from Culture Ponds

particularly, phosphorous. However, where effluents enter high energy exposed environments (open coastal sea shore) dilution and dispersion of suspended sediments are rapid. Under these conditions, the potential for negative effects on water and bottom sediment quality is low, minimizing the risk of intake water pollution.

However, if such effluents enter low energy, sheltered type of environments (i.e. in closed lagoon), the potential for exceeding the nutrient carrying capacity of the water and sediments is high. Under these conditions, combined with domestic and other agro-related nutrient inputs, anoxic conditions can develop causing prawn/fish kills in the receiving environment and loss of commercially valuable crops in the ponds.

In lake, culture of shrimps by adopting semi-intensive and intensive technologies (i.e. yield more than 5 tonnes/ha./yr.) has failed; while operations that are based outside the lake on the open coast have succeeded in Thailand.

It appears conditions in the lake are too sheltered to support large numbers of shrimp culture operations. Knowledge related to shrimp farm development gained in Songkhla, Thailand indicates that intensive or semi-intensive farming should not be carried out in Chilka.

On Lagoon Ecology

Extensive shrimp farming in the lagoon area presently utilise an estimated 25% of the lake margin strip, which normally dries during non-monsoon season. Some impingement of culture area on the permanent lake waters also occurs. As a result, the use of these areas by wild fish populations of the lagoon is eliminated. Such removal of productive areas of the lagoon is believed to have impacts on lagoon fish populations and consequently on the capture fisheries.

It is therefore, necessary to find out scientific basis about the limits to how much of this habitat can be used for prawn farming without negative impacts on wild fish population of the lagoon.

Prawn farming in Chilka is now causing conflicts with ongoing traditional fisheries and other water uses (e.g. water transportation, animal grazing etc). Expanding prawn farming in the lagoon will change the present land uses and the conflicts will intensify.

The chemistry of culture media alters due to shrimp farming, mainly by increasing ammonia concentrations and decreasing dissolved oxygen levels. Negative effects could be anticipated in areas of limited flushing or concentrated shrimp culturing activities. This could further stimulate aquatic plant growth and if discharges are large, potentially reduce fish production.

Effluents from shrimp farms is likely to have an effect on lake's phytoplankton standing stock and productivity. Additional nutrient input from shrimp farms in the lagoon could have excessive plant growth and detrimental local environmental effects.

Solid wastes in the form of shrimp faeces and unutilised feed generally deposited in the bottom of the shrimp ponds, some of which could be mixed with the adjacent lagoon water environment, resulting in physical and chemical changes to the natural sediments including increased sediment oxygen consumption and higher concentrations of total volatile solids, total organic carbon, sulphides, nitrogenous compounds and phosphates.

In net-pen culture the visible accumulation of solids and alteration of sediment chemistry normally extends approximately 30 meters from the source. Accumulation of organic-rich sediments occurs around net-pen culture unless strong currents are present. The extent of accumulation depends mainly upon water depth and current velocity.

The dissolved oxygen will be lowered near shrimp farms due to accumulation of organic rich sediments causing changes in the infaunal invertebrate community, crustaceans, moluscs, insects and other pollution sensitive arthropods will perish from the area. Changes in benthic community near shrimp farm and net pen enclosures with significant suspended sediment loads and heavy organic build up even upto total smoothering of benthic macroinvertebrates is possible.

Fish and other mobile fauna, such as crabs living on or above the bottom surface will be in higher densities around net pen enclosures, because of the availability of unutilised food fed to the cultured stocks and the abundant macro-invertebrates growing on net pen structures. Drainage of large volumes of shrimp farm effluents into the lake system would result in over-stimulation of aquatic plants and the enhancements of non-desirable lake varieties, such as minnows and pollution tolerant aquatic invertebrates.

Shrimp pond dyking in lake margins and construction of net pen-enclosures will influence tidal exchange in the lake.

In shrimp ponds, out breaks of disease is often associated with some form of stress, such as, over crowding, under nourishment, poor water quality or physical damage associated handling and confinements and the disease is frequently experienced in cultured ponds than wild stock. The organic enriched bottom sediments of aquaculture ponds could promote the growth of disease species pathogenic to humans and shell-fish collected in the vicinity of shrimp outfalls could serve as a route for infection to humans, when consumed.

Criteria for Aquaculture in Chilka

Any aquaculture options that do not satisfy the following criteria are not considered technically feasible for Chilka lake as part of a longterm production strategy.

A. Physical/Chemical

Salinity which will generally range from 5 to 27 ppt. with a mean salinity from 18 to 22 ppt. During monsoon events there may be periods when the lagoon is under freshwater influence. Some lagoon areas will be under freshwater influence continuously.

Oxygen will fluctuate in the range from 3 to 8 ppm; although with current trends towards eutrophication, there may be algal bloom events which depress levels below 3 ppm.

Temperature will range between 28 to 35 degrees celsius.

Habitat for the aquaculture species under consideration has to be compatible with that of a brackishwater lagoon habitat.

Currents will likely range from 0 to 50 cm/s (up to 1 knot) but tend to be negligible in the shallower areas, where aquaculture practices are likely to occur.

Enrichment is a problem in the system now and therefore any net nutrient additions from the aquaculture activity will have to be low.

Transparency as a measure of water clarity will be low as an enriched system is usually characterized by medium to high turbidity (i.e. low transparency).

B. Biological

Life Cycle for a species to be considered for aquaculture in this area, should be known so that options for artificial rearing of future generations can be assessed as possible probability. Also the possible impacts of escapement on the natural system must be known.

Broodstock/seed must be available initially from reliable commercial or natural sources in sufficient quantities and at an economically feasible price.

Early Life Stages required for successive crops will ultimately be required in greater quantities if a viable aquaculture industry develops for a specific target species. The rearing technologies for the required seed or larval stages should be available.

C. Other Concerns

Markets should be identified as either domestic or export with an evaluation of expected price received at the point of culture operation as well as additional processing, transport, and packaging costs. Some information on past market performance and anticipated market capacity should be known.

Return to the producer should be calculable and provide sufficient margin to cover expected rearing costs.

Technical Constraints should not be an impediment to the culture activity at any stage from the acquisition of the initial life stages through to the final markets.

Human Pathogens can be accumulated in various aquatic food chains and species. Given in high degree of human influence in the area, any aquaculture species implicated in the possible amplification of human pathogens or industrial toxicants should be avoided.

The applications of these initial screening criteria will limit the options for aquaculture in Chilka lagoon to those types that are both biologically suited for the area and have an acceptable degree of risk from an economical and social prospective.

Proposals to Protect Ecology of the Lagoon

Renewable fisheries resources within Chilka lagoon provid a sustainable protein source and economic commodity of critical habitats and populations of the lagoon can be protected and maintained. Political, social, economic and technical factors all pose significant constraints for effective fisheries management in Chilka lagoon.

Chilka lagoon without enhancement of salinity level from the present level, will undergo a steady transformation from the current state to a system that is increasingly less productive without some form of increased circulation, the rising nutrient load from agriculture and urban activities will continue to stimulate aquatic macrophyte production. An increase in plant production in the water column will increase sedimentation rates as well as contributing increase amounts of dead vegetation, which will accumulate on the bottom. The lake system will become stagnant in areas of plant growth as the oxygen in water column is used by aerobic bacteria. Ultimately the bottom sediments become anoxic which results in the production of hydrogen sulphide (H_2S).

This process will affect aquaculture operations as well as natural fish populations of the lagoon. Hydrogen sulphide is toxic to fish and crustaceans. Dystrophic events (mortalities) would coincide with periods of low circulation, which naturally occur towards the end of the dry season. This would be a period when water would be required for culture operations, which would tend to be getting ready for harvesting before the onset of the monsoon season.

The culture of herbivorous fish (Grass carp) is suggested as one possible means of mitigating these macrophyte growth. Any macrophytes removed as fish feed would obviously reduce the biomas that would ultimately be available for decay in the lake.

Without enhanced circulation and salinity, any aquaculture activity would likely be at increasing risk of these dystrophic events. Earthen ponds or pen enclosures isolated from the direct lake circulation will risk the water management measures leading to lower production level. The successful aquaculture activities of Chilka lagoon, will therefore, be to manage larger areas, preferably the lagoon as a whole, without creating any isolated areas and obstructions in the lagoon system with lower rates of production.

The Chilka lake itself can be stocked with prawn seeds like the Seto Island of Japan, to get more catch of prawn. As the lake is shallow and eutrophic in nature having optimal salinity range not exceeding 20 ppt, it provides better conditions for faster growth of prawns than in sea. The escape of stocked seeds is highly restricted as the lake is connected to sea with narrow channel. Increased freshwater influence in the Chilka lagoon will preclude many of the organisms that are currently found in the brackishwater system. This will have impact on both existing aquaculture practices, as well as, many of the brackishwater and salt-water species represented in lagoon's flora and fauna. For the impact that these changes would have on the local and regional economics it is advisable to intervene by enhancing the circulation of sea-water to the lake system to project the estuarine character of Chilka ecosystem with all its genetic diversity, optimization of salinity condition of the lagoon for improving fishery potential which will have significant bearing on the socio-economic condition of large numbers of poor fishermen living in and around the lake.

Desilting of the lagoon and especially of Magarmukh and Palur canal by dredging, will ensure the flow in and out of Chilka to Bay of Bengal and help in restoring the salinity gradient of the lagoon.

Steps should also be taken to augment fishery potential by higher recruitment of prawn and fish juveniles from adjacent Bay of Bengal to the lake.

Restoration and optimization of the salinity gradients of the lagoon can be achieved by—

(1) Dredging of Magarmukh area and the connecting channel between the lake and adjacent Bay of Bengal.

(2) Establishing a new inlet.

Mechanical intervention would be most effective if the lagoon system can be regulated to maintain salinities in the range of 5 to 27 ppt., typical for brackishwater systems and will support the existing biota. The mixing process of the lake will assist in distributing saline water from the point of entry to all areas of the lake. This will not only increase the exchange of salt water between the lake and the Ocean, but also some alternation to the lake shorelines to improve natural circulation patterns and to improve sediment removal from the lake system and further enhance mixing processes within the lagoon.

With properly enhanced system, many of the problems currently encountered in the lagoon system would be mitigated. Sedimentation from local agricultural practices would be countered by increased current speeds and improved sediment transport out of the lagoon system. Circulation canals could be maintained by dredging, which would reduce the burden placed on the shallow littoral areas to accommodate increased sediment deposition. The productive littoral intertidal areas of the lagoon can be left and conserved without any intervention for grazing of juveniles of fish and crustaceans, thus recruited from the Bay of Bengal.

The nutrient loading rate, which is currently resulting eutrophication of Chilka would decrease in direct relation to the enhanced rate of exchange of low nutrient rich oceanic water.

In the enhanced system Chilka would maintain optimum brackishwater conditions with higher salinities over more of the lagoon areas than that of present level. The northern part of the lake and other areas receiving fresh water input would provide localised freshwater lens areas and shallow littoral areas under freshwater influence. In addition, the lagoon would be subjected to seasonal flushing with freshwater during heavy monsoon rainfall generally occurring from June to September each year. Salinity regions will be stable over most of the lake areas other than during freshwater events associated with the monsoon season.

This stabilized brackishwater environment, after initial enhancement efforts in Chilka lagoon will offer opportunities strongly in favour of production of marine shrimps (*Penaeus monodon & Penaeus indicus*) and fish (mullets, seabass, perches and other oceanic varieties) with the following advantages.

(i) they are of high economic values.

(ii) there are existing marketing infrastructure and contacts.

(iii) seed for autostocking and supplmentary stocking available locally.

(iv) technical know-how available locally.

A shore-line development concept for developing upland areas (e.g. old lake), now used for paddy cultivation and animal grazing, can be efficiently used for aquaculture by excavating ponds in old lake beds behind the present shore-lines, particularly to the north of Chilka lake culturing the following species of fish and prawn for higher economic returns.

1. Freshwater prawn (*Macrobrachium rosenbergii*): These prawn can be cultured in low salinities using similar techniques to whose currently used in extensive rearing of marine shrimp (primarily *Penaeus monodon*). This species exhibits some aggressive territorial behaviour under high intensity rearing conditions but this should not be a problem at the production levels recommended for Chilka Lake. Freshwater prawn are currently cultured in India (Pathak, 1989). Seed stock availability in the Chilka areas will be required to support a viable culture industry. Export market prices in Thailand for freshwater prawns have been 40 per cent to 70 per cent lower than marine shrimp (New, 1990). Given India's complicated and efficient export marketing infrastructure (Pathak 1989), it is likely that a freshwater prawn culture industry in Chilka Lake would be dependent on domestic markets in the near term for economic viability.

2. Grass carp (*Ctenopharyngodon idella*): This carp species has been reared extensively in India, especially as one species in carp polyculture systems (Pathak, 1989). This species can be cultured in low salinity ponds and has been suggested as a candidate for culture in Chilka Lake to achieve the two-fold objectives of fin fish production on the one hand and control of the nuisance macrophytes (*Potamogeton* sp.) on the other hand. Seed would likely be available given the current well developed carp culture programme in India (ADCP,1988).

3. Grey Mullet (*Mugil cephalus*): This species is amenable to culturing in brackish to freshwater conditions. This species is herbivorous and is currently used extensively in India in polyculture (fin fish/shrimp) culture operations. This approach could be extended to mullet/prawn culture operations in the existing Chilka Lake culture purpose. As per the grass carp options, grey mullet rearing operations could utilize some of the macrophyte production in the lake as a food source. Recruitment of juveniles from broodstock is a well documented process and should be available to provide seed for Chilka Lake operators.

4. Milk fish (*Chanos chanos*): These fish are also herbivorous and are currently being raised in fresh and saltwater ponds in many Asian countries including India. This species occurs naturally in Chilka Lake and constitutes a small component of the commercial catch. The rearing techniques are well known and this species has also been shown to perform well in polyculture operations, when raised in conjuction with shrimp (NACA, 1987).

5. Pearlspot (*Etroplus suratensis*): These fish currently constitute the most numerous fish species in Chilka Lake. They are amenable to rearing in both freshwater and brackishwater conditions. They are herbivorous eating both aquatic plants and plankton. Being as numerous as they are in the lake, it would require little in the way of technology to start a program of pond rearing using seed captured in the wild. As per the other herbivorous fish, the pearl spot may possibly be reared on a diet of macrophytes originating in the lake.

Fisheries and Wild Life Management

Awareness among residents in and around Chilka lake and general public are to be promoted about the lake system and their participation in the management process have to be ensured.

A fisheries management programme for protecting important fish habitats, fish stocks and regulating fishing activities have to be strictly enforced. For that a community based resource conservation and development with extensive local village participation and responsibility is necessary.

Fisheries management within Chilka Lake occurs primarily through the allocation of leases. There are few fishing regulations, and limited capabilities for enforcement. While this system operated reasonably effectively in the past, the increased demands placed upon the resource by an ever-increasing number of participants in the lagoon fishery have caused over-exploitation and inefficient utilization of the resource. In future, it will be necessary to develop effective mechanisms which control total fishing effort, and serve to allocate fishery resources in an equitable fashion. Any such mechanisms need to be carefully and sensitively designed, in view of the dependence of many local communities and fisherfolk upon Chilka Lake fisheries resources. Potential overharvesting of several marine species (including prawns) within the lagoon is offset by the recruitment of juveniles from neighbouring marine areas within the Bay of Bengal. It is critical for the sustained operation of the Chilka Lake fishery, that such marine areas are maintained as reserves and protected from fisheries exploitation in future. Within Chilka Lake there are currently no conservation zones, where fish stocks are protected from exploitation (with the exception of the Palur Canal and the lagoon mouth). Such conservation zones need to be established in areas, where fish spawning is localized.

Obstruction of Palur Canal with the help of Bamboo and Net

Available data are inadequate to convincingly demonstrate, whether over-fishing is significantly affecting Chilka Lake fish stocks. Qualitative assessments (low modal size of recruits) and the relatively large number of fishermen involved in the fishery suggest the possibility of over-fishing and an inefficient utilization of the resource. There is a requirement to obtain better fisheries data for monitoring trends in Chilka Lake fish populations.

Although Chilka Lake is one of India's designated Ramsar sites for waterfowl, bird management programs are non-existent, and the populations are susceptible to direct human interference, as well as habitat-related changes. There is a pressing need to ensure that the birds receive adequate protection, through designation and enforcement of bird conservation areas. There are a set of enhancement-related activities that should be undertaken to create appropriate habitat features, which promote bird population growth and rare species survival.

Available information on Chilka Lake dolphins is virtually non-existent, necessitating the development of management strategies based on general principles. It is desirable to maintain dolphin populations within Chilka because they are part of India's natural history legacy. Dolphin populations also provide a potential attraction for wildlife viewing.

Fisheries Management with "No Change" Condition

In the past fisheries management within Chilka lagoon was primarily through the allocation of leases of fisheries sources of the lake primarily for capture fisheries (except Jano fisheries, where impoundment of fishes and prawns for short period i.e. for 2 to 4 months have been made so as to grow and catch in intervals).

Since 1985, higher prices for prawns has led to the commercialization of prawn farming in Chilka lagoon, which has made the non-traditional fishermen, particularly the farmers living in and around the lagoon, to enter into prawn culture. This resulted more and more illegal occupation of land and water in and around the lake for prawn farming.

Aquaculture activities in Chilka lake are currently predominated by extensive pond culture of black tiger prawn (*Penaeus monodon*). Optimum salinities for this species are in the range of 15 to 30 ppt. (Kungvankij, 1986). They are raised in earthen pond located around the lake margins and to some extent in the shallow of the lake. Water exchange to the ponds is intermittent during high water periods. Because of the relatively low stocking densities in extensive culture, only limited intermitted water exchange is required. If the salinity of the water is low at the time of pond filling, salt is leached from the soil under the pond and the water salinity in the pond increases.

As the ambient salinity of Chilka Lake becomes less saline (i.e.) greater freshwater influence) the salt in the soils will effectively "condition" the pond water resulting in salinities that continue to be suitable for prawn rearing. Eventually the salt in the soils will leach out and their salinity conditioning capacity will be reduced. If freshwater or low salinity conditions persist in the lake, the soils will be leached to the point where optimum salinities for prawn culture will not be met in the culture ponds. This process will likely occur over a 2 to 5 year period (K. Wiley, pers Comm.)

Reduced salinities in the lake system will not have an immediate impact on the existing shrimp culture industry. Rather the success of the various farming ventures will start to diminish with reduced survivals and growth as the optimum salinity levels are gradually not met. The operations under the greatest freshwater influence will be effected first but, given the high degree of mixing that occurs in the lake system, all operations will be affected within a similar time frame. The rate of water exchange required is a function of biomas in the ponds and so the more intensive operations will be negatively affected sooner than the low biomas extensive operations.

Operations for semi-intensive prawn culture would be very susceptible to impacts from reduced salinities. Increased freshwater influence in the lake may preclude the growing of many of the organisms that are currently found in the brackishwater system. Only those organisms adaptable to near freshwater conditions (0 to 10 ppt.) would be considered as candidates for aquaculture in an unenhanced system.

Without saltwater enhancement, the culture species that will provide the best rates of return for the farming community over the longer term will be those species that are amenable to rearing in salinities ranging from 0 to 10 ppt. This will likely be the seasonal variation in salinity given the continued freshwater influence modified by the residual salinity contained in the soils underlying the lagoon system.

Culture species suitability for Chilka Lake will change as the system evolves towards increased freshwater influence.

With the advent of prawn culture technology in early eighties and to protect the Chilka lagoon from ecological hazards, conflict and litigation in the present "no change" condition, the culture and capture fisheries together with the areas which will be considered for culture and capture fisheries in Chilka lake with minimum ecological disbalance can be defined as follows:

Capture Fishery

Capture sources in Chilka mean the traditional sources like Jano, Dian, Uthapani and Bahanis, but shall not include:

(a) the areas earmarked for culture of fish and prawn;
(b) the areas conserved for migration of fish and prawn including their juveniles;
(c) the areas earmarked as nursery and grazing grounds of juveniles of prawn and fish;
(d) the areas declared as birds and wildlife sanctuaries in the lake;
(e) the areas earmarked as migratory path of fish and prawn including their juveniles;
(f) natural navigation channels of the lake, from, where the aquatic produce (fish & prawn) can not be totally harvested/taken out at any point of time.

Culture Fishery

The culture sources will include, the controllable confined areas between inter-tidal zone of Chilka lake, where the aquatic animals (fish & prawn) can be held up for certain time in order to fatten them to grow, and the produce from the area can be totally harvested at any point of time, but shall not include:

(a) the areas conserved for migration and nursery and grazing grounds of fish and prawn including their juveniles;
(b) the areas declared as bird and wildlife sanctuaries in the lake; and
(c) areas earmarked as migratory path and navigational channels of the lake.

Explanation

The inter-tidal zone of Chilka means the area between high and low water mark of the lake in patches (not contiguous) excluding the areas earmarked for nursery and grazing ground of fish and prawn and their juveniles, migratory path of fish and prawn including their juveniles, natural navigation channel of the lake, the area declared as sanctuaries for birds and wildlife, which get fully dried and exposed to solar radiation, at least some part of the year so as to have a natural disinfected and easily biodegradable top soil before starting the next culture operation.

All capture sources may be leased out to the Central Fishermen Coop. Marketing Society, Balugaon who in turn shall sub-lease them to the affiliated primary societies. The capture fishery sources may be divided into convenient operational size in the interest of better management. In no case, the operational size shall exceed 1000 acres. The annual lease value of each capture source (prawn and non-prawn) shall be fixed at by adding 10 per cent to the preceding years lease value.

Each such primary society shall be given a culture fishery source of not more than 100 acres at the upset price of Rs.800 per acre per annum with 10 per cent increase in successive years.

Rest of the culture sources can be leased out to the societies/organisations formed by the inhabitants of neighbouring villages who are not members of the primary societies of fishermen.

The period of lease shall be for three years beginning from the 1st day of January every year with 10% increase in lease value every year.

On expiry of the period of lease the right of the lesee to operate the source shall automatically cease.

The settlement of fishery sources as indicated above shall be subjected to the following stipulations.

(a) In the culture fishery only extensive prawn culture will be allowed without addition of any organic and/or inorganic fertilizer, supplementary feed, chemicals, pesticides and antibiotics at any time.

(b) The outer channel, Palur canal and navigational channels of the lake together with conserved areas earmarked for nursery and grazing grounds of fish and prawn juveniles shall not be obstructed in any way by fixing any device, such as earthern embankment, netting of any sorts, fixing pen materials etc.

(c) The radius of one kilometre area in the lake mouth and at the opening of outer channel into the lake, Rushikulya mouth and at the opening of Palur canal into the lake shall be kept free from any obstruction as indicated above.

(d) The areas for prawn culture and conservation for nursery and grazing grounds of the prawn and fish juveniles already identified shall be demarcated in consultation with noted fishery scientists and ecologists and the identification of such areas will be reviewed and monitored at an interval of every three years keeping the above principles in view before leasing the areas afresh.

(e) The lesee shall operate the source directly and shall under no circumstances sub-let/sub-lease/permit the sources to any body and in the event of such sub-letting the source lease shall be forfeited and the lessor shall be competent to enter upon, take over possession of the fishery sources and lease it out to the others. In such circumstances the lesee shall not be entitled to any compensation and lessor in addition to taking over the sources shall realise the full lease value if not realised till then.

(f) The lesser (or any officers duly authorised by him) shall retain the right to enter into the fishery sources for the purpose of inspection, survey, measurement, enforcement of lease condition or any other purpose as deemed necessary. The lesee shall be bound to assist the lessor or any of his (subordinate Officers duly authorised by him) to take up such survey monitoring etc.

(g) The lesee shall always be liable to make payment of lease amount by the date as may be determined by the lessor failing which he shall be liable to pay interest @12% per annum from the date of such default over and above the outstanding lease amount.

(h) No mud wall embankment shall be erected to enclose culturable source, thus identified but with pen netting materials not below 3 mm. square mesh pen wall supported by bamboo and wooden poles so as to allow free exchange of water between the culturable area and the main lake.

(i) The extent of fishery source so determined shall be in conformity with that of the map if the source maintained by Collectors of Puri, Khurda and Ganjam districts and the Tahasildars concerned. For this purpose boundary marks shall be fixed on the spot to identify the sources and distinguish one source from the other which are to be indicated in the map prepared for the purpose.

(j) Not more than 5 (five) leader lines may be laid in a prawn source and distance between 2 (two) leader lines may not be less than half a kilometre.

(k) Leader line (Khandas) may be laid only on one side of the source extending from the land mass upto the main channel/Bahar Chilka. But in no case length of the leader line should exceed 250 m. indicating the maximum distance between the land mass upto the midstream of Bahar Chilka and if the distance is less than 250 m. then the length of the leader line may be limited to that extent. In other words, if the distance of the leader line from the land mass upto the midstream of Bahar Chilka will be limited to 250 m. only.

(l) The leader lines shall be removed after the prawn season is over, but the lesee society shall have the right to catch fish by other means till the lease period expires subject to the condition as above.

(m) The period of operation for different types of sources shall be as follows subject to the restrictions imposed above.

(i)	Bahani	Throughout the year.
(ii)	Jano	From September to January.

(iii)	Uthapani	From July to September.
(iv)	Dian	From July to October.
(v)	Prawn	From February to July.

Size of traps and nets etc. used as devices for catching fish shall be as follows:

1. Bazza — Its height, length and breadth shall not exceed 32", 21" and 8" respectively.
2. Dhaudi — Its height, length and breadth shall not exceed 12", 36" and 12" respectively.

(n) Collectors and the concerned Tahasildars shall have the power to seize boats/ nets/ bazza/ dhaudi etc. involving clandestine catching of fish. On every such seizure minimum penalty of Rs.1000 (Rupees one thousand only) shall be imposed on boats including out-board engine fitted boats having hull length of less than 25' and Rs.2,500 (Rupees two thousand and five hundred only) for boats having hull length of more than 25 ft. and Rs.5,000 (Rupees five thousand only) for mechanised boats such as trawlers.

(o) Violation of any one or more of the aforesaid conditions shall be sufficient for termination of lease by the Collectors concerned.

(p) The machinery for enforcement of this lease principles shall consist of one platoon of Armed Force with an Executive Magistrate who shall move frequently in the lake and shall be answerable to Collectors of the concerned district.

(q) The license fee of Rs.50 (Rupees fifty only) each per annum shall be paid by all boats which are not larger than full length of 25ft. and Rs.100 (Rupees one hundred only) for boats larger than this size. They shall be allotted serial numbers, which would prominently be displayed in red colour over white both on the stem end on the stern.

(r) The conditions of the lease deed that shall be executed between the Collector and the Cooperative Marketing Federation, Bhubaneswar shall form part of the lease deed to be executed between the Primary Fishermen Coop. Society and the Cooperative Marketing Federation, Bhubaneswar.

(s) Restrictions as would be imposed by the Fisheries and Animal Resources Development Deptt. under the Orissa Marine Fishing Act, 1981 shall be followed scrupulously, which relates to conservation of Chilka lake fisheries as under—

The State Government do hereby regulate fishing in the Palur canal and Chilka in the following manner.

(i) Fishing by any method is prohibited in Palur canal throughout the year.

(ii) Capture of Khainga, Kabla Bhekti below 150 (one hundred fifty) mm. size and prawns like Bagda and Chapra (Kalipoli) varieties below 100 (one hundred) mm. size by any means is prohibited throughout the year in Chilka lake.

(iii) Fishing is completely prohibited in the outer channel of Chilka lake during the months of December and January.

(iv) Prawns which are not growing above two inches as specified below can be caught in the outer channel of the Chilka lake during the months of February and March.

Descriptions of prawns Scientific name	Local name (in Oriya)
1. *Metapenaeus monoceros*	MORODA
2. *Metapenaeus dobsonii*	PANU
3. *Metapenaeus affinis*	GHASUA

(v) No fishing by means of net shall be allowed in the outer channel throughout the year. Fishing, however, by other means like Flianda, Thatta and Baza can be undertaken in the outer channel. But no Khanda shall enter into the deeper zone of the outer channel.

The areas of outer channel and Palur canal are as stated below:

Outer channel: Starting from tip of Retamati (Satpada) upto mouth of Chilka lake.
Distance 25 kilometres within
Latitudes 19.39′00 „
19.44′00 „
Longitude 85.26′00 „
85.38′00 „

Palur canal: Starting from Rambha Bay upto the mouth of River Rushikulya.
Distance 15 kilometres within
Latitude 19.22′45 „
19.29′30 „
Longitude 85.04′45 „
85.08′30 „

The State Govt. do hereby prohibit fishing of prawn larva (PL-20) and Juveniles (30 mm to 50 mm size) from (a) Krushna prasad up to 2 kms. north of Chilka Lake mouth in Puri District (b) 2 km. of South of Chilka Lake mouth covering Rushikulya estuary and Palur canal in Ganjam District.

References

1. B.O.B.P.	The Jano Fishery and the Fisheries Regulation of the Chilka Lagoon, Orissa, India. AMFS—Report by UIF Nermark, Gothenburg, May 1989.
2. Biswas, K.P.	On the Production of Fish and Crustaceans from the Chilka with Environmental Changes and its Impact on Fishermen of the Lagoon in Orissa State.
3. Biswas, K.P.	On the Trend of Fishery of Chilka Lagoon.
4. Biswas, K.P.	On the occurrence and Exploitation of Fish and Prawn from Chilka Lagoon: A Natural Cultural Ecosystem of Orissa State.
5. Biswas,K.P.	Environmental Management in Coastal Area—Marine Fisheries.
6. C.I.D.A.	Chilka Lake Project: Feasibility and Design Report.
7. Das, G.S.	A Study of the Problems of Fishermen Cooperatives in Chilka, Dep. of Economics, G.M. College, Sambalpur India, March 1976.
8. Das M.K. & Samal R.	Environmental Survey of Chilka, ORSAC, Bhubaneswar.
9. Deptt. of Science Technology & Environment.	Base-line information on Chilka, Bhubaneswar, India, 1988.
10. Directorate of Fisheries, Govt. of Orissa, 1991.	Hand book on Fisheries Statistics, Orissa.
11. Jhingran V.G. & Natrajan A.V.	Study of the Fisheries and Fish Populations of the Chilka Lake during the period 1957-1965, Inland Fish Soc. India, Dec. 1969.
12. Katre D.B. & Das Narayan G.R.	Deptt. of Atomic Energy, Atomic Energy Division (SZ), Berhampur.
13. Mohapatra P.	Our progress, Dept. Fish., Cuttack, Orissa, 753007, Fish Bull, Vol-I Jan 1986.
14. Mohanty, S.K.	Some observations on the physico-chemical features of the outer channel of the Chilka lake during 1971-73. Bull Dept. Mar. Sci., Univ. Cochin, 1975, VII, 1, 69-89.
15. Mohanty, S.K.	The breeding of economic fishes of the Chilka Lake—A review , Bull Dept. Mar. Sci., Univ. Cochin, VII,3. 1975, 543-559.

16. Mohanty, S.K. — Some Measures of Conservation and Development of Fisheries Resources of the Chilka Lagoon, BFDA, Puri, India, 1988.

17. Mohanty, S.K. — Small-scale Brackishwater shrimp-culture in the Chilka Lagoon and its ecological and socio-economical significance, Presents at the seminar on Conservation and Management of the Chilka Lagoon Bhubaneswar, Jan 20-22, 1988.

18. Mohanty S.K. — National utilization to Brackishwater resources of Chilka Lagoon, Deptt. of Fisheries, Orissa, Cuttack.

19. Murty, A.S.N. — Marine Science Deptt., Berhampur University, Berhampur.

20. Mishra, P.M. — Fisheries Development of Chilka Lake, Deptt. of Marine Sciences, Govt. of Orissa, Berhampur.

21. Panigrahi, G. — Vegetation and flora of Chilka Lagoon Botanical Survey of India, Calcutta.

22. Rajyalakshmi T. & Ravichandran P. — Pen and cage culture management of Chilka Lagoon, Presents at the seminar on conservation and management of Chilka Lagoon, Bhubaneswar, Jan 20-22, 1988.

23. Ravichandran Patel — On the Growth of *Panaeus monodon* in pens in Chilka Lagoon and its Ecological Significance, Puri Research Centre of Centr. Inl. Fish. Res. Inst., Orissa, India, 1987.

24. Rayachaudhuri, B. — The Fishing Communities of Chilka: A study in their Ecosystem, J.Indian Anthrop. Soc 101-112, Calcutta, India 1977.

25. Rama Rao, K.B. & Nageswar Rao C.A. — Fauna of the Chilka Lagoon, Estuarine Biological and Zoological Survey of India, Berhampur.

26. Rao, D.P., Subramanium S.K. & Sudarshan R. — Geographical analysis on Chilka Lake and Adjoining area of using Remote Sensing Data, Indian Institute of Remote Sensing, National Remote Sensing Agency, Dehradun.

27. Sahoo B.N. — The Chilka Lake is in danger, Deptt. of Agronomy, O.U.A.T, Bhubaneswar, Orissa.

Index